I. J. Douglas HNC(Mech.Eng.)
Principal Teacher of Technical Subjects,
Madras College, St. Andrews.

J. D. G. Crichton C.Eng. MRINA
George Heriot's School, Edinburgh.

SI Engineering Mechanics

Oliver & Boyd

Oliver & Boyd
Tweeddale Court
14 High Street
Edinburgh EH1 1YL
A Division of Longman Group Limited

ISBN 0 05 002319 5

Set in 10/12pt Plantin 110 and
the Univers Series and
printed in Great Britain by
T. & A. Constable Ltd., Edinburgh

The drawing on the cover is reproduced by
courtesy of the Armfield Engineering Co. Ltd.

Contents

Preface

The object of this book is to present the basic principles of applied mechanics in a concise but clear form. To this end no experimental work has been included, and descriptive material has been kept to a minimum. After the introduction, each chapter consists of a short account of the relevant theory followed by a comprehensive series of worked examples and is concluded by a set of problems to be solved by the student.

Though primarily intended for pupils taking the Scottish Certificate of Education examination in Applied Mechanics (O Grade), the text has been extended to cover the applied mechanics sections of the Joint Matriculation Board examination in Engineering Science (O Level) and the G1 and G2 years of the General Course in Engineering. It should prove useful both in secondary schools and in the earlier years of technical college courses.

Since the Système International (SI) is comparatively new to many people in this country we felt that the introductory chapter was necessary and would give an understanding of the basic units involved in the system. All detail complies with the following publications which should be consulted for further information:—

1 THE USE OF SI UNITS British Standards Institution: PD5886.
2 CHANGING TO THE METRIC SYSTEM National Physical Laboratory: HMSO.
3 SYMBOLS FOR SCE EXAMINATIONS SCEEB circ. 9/20, 1967.

Practical engineering cannot be divided into rigid compartments, and we have tried throughout the text to break down the barriers between topics. Thus, for example, a machine problem may also involve moments or triangle of forces. This serves the double purpose of helping the student to adopt the correct approach right from the beginning and it also acts as a revision for the earlier topics.

1
Introduction: SI Units

The initials SI stand for Système International d'Unités; this title was given at an international conference in 1960. The system is based on metric units but, unlike the older metric systems, is a coherent system: the unit of any composite quantity is formed from the units of the separate quantities, *e.g.*

$$1 \text{ joule (work)} = 1 \text{ newton (force)} \times 1 \text{ metre (distance)}$$

This has the advantage that no awkward conversion factors are required, as was the case with the British foot-pound-second system of units. This simplifies calculation and reduces the chance of making mistakes. Mechanics, heat and electricity are brought together in SI and we no longer have, for instance, three separate units for energy. The unit of energy is the joule and it applies equally to mechanical, electrical or heat energy.

Basic Units

length—metre(m)   electric current—ampere(A)
mass—kilogramme(kg)   temperature—kelvin(K)
time—second(s)   luminous intensity—candela(cd)

Only the first three will be required in this book; the others are included for completeness. All other units are formed from these basic ones and are known as derived units.

Mass is the amount of matter or substance in a body. The mass of any body is constant and is independent of the position of that body.

Force Applied mechanics is chiefly concerned with forces and the effects of forces. It may thus be a little puzzling at first to find that force is not a basic quantity. The unit of force is derived directly from Newton's second law of motion, which may be written in the form:—

$$\text{Force} = \text{Mass} \times \text{Acceleration}$$

thus: 1 unit of force = 1 unit of mass × 1 unit of acceleration
$$= 1\,\text{kg} \times 1\,\text{m/s}^2$$

The unit of force is the newton(N) and it is that force which will give a mass of 1 kg an acceleration of $1\,\text{m/s}^2$, *i.e.*

$$1\,\text{N} = 1\frac{\text{kg}\,\text{m}}{\text{s}^2}$$

Since problems involving both forces and masses are common in mechanics, it is essential to check all equations to make sure that the units on both sides of the equation are consistent.

Force of Gravity   The force of gravity is the force with which all bodies are attracted to all other bodies. The magnitude of the force is proportional to the masses of the bodies, inversely proportional to the distances between them and is directed toward their centres. It is gravitational attraction which, for example, keeps the earth in orbit round the sun and the moon in orbit round the earth. Here we are concerned only with the earth's gravitational field.

The force of gravity varies with the distance from the centre of the earth, and since the earth is not a perfect sphere and all points are not at sea level, then there is a slight variation in the force from point to point on the surface. The force of gravity gives an acceleration to any mass falling freely, and as the force varies, then the acceleration varies. Since the variation is slight, a standard acceleration due to gravity(g) of $9 \cdot 81\,\text{m/s}^2$ has been adopted which is sufficiently accurate for most engineering purposes. A force of 1 N gives a mass of 1 kg an acceleration of $1\,\text{m/s}^2$. A force of $9 \cdot 81$ N will be required to give a mass of 1 kg an acceleration of $9 \cdot 81\,\text{m/s}^2$.

i.e. The gravitational force on a mass of 1 kg is $9 \cdot 81$ N and the gravitational force on a mass of m kg is $m \times 9 \cdot 81$ N.

The gravitational force is usually called the weight.

$$\text{Weight} = \text{m} \times \text{g}$$

Multiple and Submultiple Units

The basic units do not satisfy the requirements of every situation in that they may give answers with a large number of figures. To overcome this situation a series of multiple units may be used. In order to understand the system fully it is essential to become familiar with the index notation.

$1000 = 10 \times 10 \times 10$; this may be written as 10^3 (10 to the power 3). Similarly: $100000 = 10^5$; $1000000 = 10^6$; $\frac{1}{1000} = \frac{1}{10^3}$, which may also be written as 10^{-3} (10 to the power minus 3).

If the figures are not unity they are written as shown in the following examples:

$$500 = 5 \times 10^2; 4\,560 = 4 \cdot 56 \times 10^3$$
$$0 \cdot 05 = \tfrac{5}{100} = 5 \times 10^{-2} \; or \; 50 \times 10^{-3}$$

It is recommended that as far as possible only powers of multiples of three be used (10^6, 10^{15}, 10^{-3}, 10^{-6}, etc.).

The prefixes of the main units are shown below:

giga	G	10^9		centi	c	10^{-2}
mega	M	10^6		milli	m	10^{-3}
kilo	k	10^3		micro	μ	10^{-6}
deci	d	10^{-1}				

Values should be given in the range $0 \cdot 1$ to $1\,000$:

$11\,500\,000\,N$ should be written $11 \cdot 5\,MN$ or $11 \cdot 5 \times 10^6\,N$
$0 \cdot 005\,N$ should be written $5\,mN$ or $5 \times 10^{-3}\,N$
$15\,250\,m$ should be written $15 \cdot 25\,km$ or $15 \cdot 25 \times 10^3\,m$
$0 \cdot 000\,004\,m$ should be written $4\,\mu m$ or $4 \times 10^{-6}\,m$

In the case of mass, 1 megagramme is not $10^6 \times$ the basic unit, kilogramme, but $10^3 \times$ kilogramme. This is exceptional and arises from the fact that kilogramme and not gramme was chosen as the basic unit. Since this might lead to confusion in the early stages, 10^3 kg or the tonne (t) is used in this book and the megagramme is ignored. The tonne ($= 10^3$ kg) is a convenient unit for large masses, though it should be noted that it is not strict SI, and must be converted to kg when used in most equations.

Areas and Volumes

These may be required in square or cubic metres, though the dimensions given are in some other unit. The following examples explain the method to be adopted:

1   Find the area, in square metres, of a rectangle 5·5 km by 2·5 km

$$5·5 \, km = 5·5 \times 10^3 \, m, \quad 2·5 \, km = 2·5 \times 10^3 \, m$$

$$\text{Area} = (5·5 \times 10^3) \times (2·5 \times 10^3) \, m^2 = 13·75 \times 10^6 \, m^2$$

$$[10^3 \times 10^3 = (10 \times 10 \times 10) \times (10 \times 10 \times 10) = 10^6]$$

2   Find the area, in m^2, of a circle 20 cm in diameter

$$20 \, cm = 20 \times 10^{-2} \, m$$

$$\text{Area} = \frac{\pi}{4} d^2$$

$$= \frac{\pi}{4} (20 \times 10^{-2})^2$$

$$= \frac{\pi}{4} \times 20^2 \times (10^{-2})^2$$

$$= 314 \times 10^{-4}$$

$$\text{Area} = 31·4 \times 10^{-3} \, m^2 \text{ using preferred notation}$$

3 Find the volume, in cubic metres, of a cylinder 20 cm in diameter and 63·6 cm long.

$$V = \frac{\pi}{4} d^2 l$$

$$= \frac{\pi}{4} \times (20 \times 10^{-2})^2 \times (63·6 \times 10^{-2})$$

$$= \frac{\pi}{4} \times 400 \times 10^{-4} \times 63·6 \times 10^{-2}$$

$$= 314 \times 63·6 \times 10^{-6}$$

$$= 20 \, 000 \times 10^{-6}$$

$$= 20 \times 10^3 \times 10^{-6}$$

$$\text{Volume} = 20 \times 10^{-3} \, m^3$$

The litre (l) is sometimes used for volumes of fluids. 1 litre = 1 cubic decimetre, or 1000 litres = 1 m^3.

Most units are named after important scientists, and when

written out in full they start with a small letter. If however they are abbreviated they are given a capital letter, and never have an 's' to denote the plural, *thus*:

$$600 \text{ watts } or \text{ } 600 \text{ W}$$

When solving problems a definite technique should be adopted. Wherever possible (and there are very few cases where it is not possible) a sketch should be drawn, showing as much information as possible. All the necessary information should be extracted from the question and put at the head of the page. The next stage is to look at the information and ensure that it is in the correct units for solution: *e.g.* given 52 km/h the solution will usually be required as metres/second:

$$52 \text{ km/h} = \frac{52 \times 1000 \text{ m}}{60 \times 60 \text{ s}} = 14 \cdot 45 \text{ m/s}$$

$$i.e. \quad 1 \text{ km/h} = \frac{10}{36} \text{ m/s}$$

Before any figures are substituted, the basic equation or equations to be used must be quoted either in symbolic form or in words. After the calculation has been completed, the answer should be given in statement form, care being taken that the unit is stated as well as the quantity. Work from first principles whenever possible.

NOTE: The diagrams in this book are not drawn to scale. Where problems are to be solved by graphical methods, convenient scales are suggested.

2
Forces:Graphical Methods

Force A force is that which changes, or tends to change, the state of rest or uniform linear motion of a rigid body.

Unit of Force The unit of force is the newton (N).

Specification of a Force Before a force can be completely specified three aspects should be considered:
a) The *magnitude* of the force.
b) The *direction or sense* in which the line of action of the force acts.
c) The *point of application* through which the line of action of the force acts.

Vector Quantity A vector quantity possesses both *magnitude* and *direction*.
Force, which satisfies the above two conditions, is therefore a vector quantity.

Scalar Quantity A scalar quantity possesses *magnitude* only.

Equilibrium When a body is at *rest*, or *moving with uniform motion* in a straight line, it is said to be in equilibrium (balanced).

Action and Reaction To every action there is an equal but opposite reaction required to obtain equilibrium (Figure 2.1).

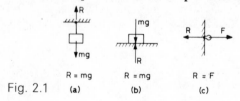

Fig. 2.1   (a)   (b)   (c)

If two, three or more forces are not in equilibrium, then by various means each set of forces can be replaced by a single unbalanced force called the *Resultant* which can be balanced by a single force called the *Equilibrant* (Figure 2.2).

The Resultant is that single force which can replace a number of forces and have the same effect.

The Equilibrant is that single force which can balance the resultant force, or a system of forces.
Note that the Resultant and Equilibrant are equal in magnitude but opposite in direction.

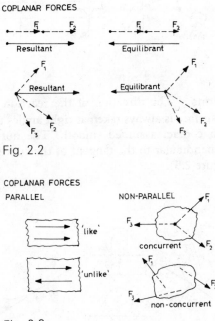

Fig. 2.2

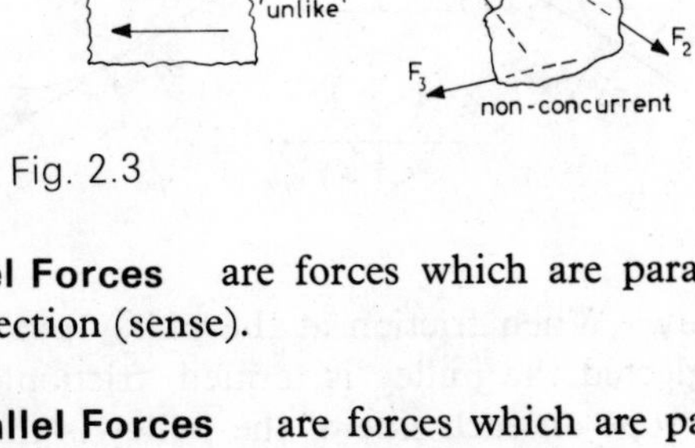

Fig. 2.3

Like Parallel Forces are forces which are parallel and have the same direction (sense).

Unlike Parallel Forces are forces which are parallel but act in opposite directions.

Concurrent Forces   are forces whose lines of action meet at the same point.

Coplanar Forces are forces which act in the same plane.

Reactions at 'Smooth' Surfaces When a surface is considered smooth (friction neglected) the line of action of the reaction will be at right angles to the smooth surface.

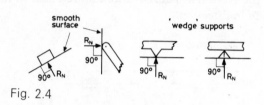

Fig. 2.4

Roller Reactions The direction of the reaction between the roller and the surface is always taken at right angles to the surface the surfaces in contact assumed smooth. The normal reaction R_N will be perpendicular to the tangent of the curve at the point of contact (Figure 2.5).

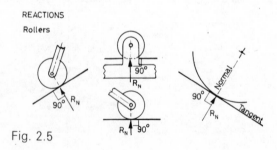

Fig. 2.5

Smooth Pulley When friction at the pulley pin and bearing bracket is neglected the pulley is termed 'frictionless' and the tension in the rope on each side of the pulley is the same. The mass of the pulley and rope is also usually neglected (Figure 2.6).

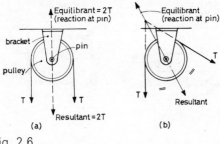

Fig. 2.6

The Resultant and Equilibrant (reaction at the pin) can be obtained by:
a) Parallel forces (Figure 2.6a)
b) Parallelogram or Triangle of Forces (Figure 2.6b).

Representation of a Force   A force can be represented on paper by means of a straight line (Figure 2.7):
a) *Magnitude:* the length of the line drawn to a suitable scale represents the magnitude of the force.
b) *Direction or Sense:* an arrowhead placed on the line indicates the direction (sense) of the force.
c) *Point of Application:* one end of the line represents the point of application of the force.

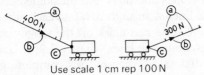

Use scale 1 cm rep 100 N

Fig. 2.7

Method of finding the resultant of parallel forces
The resultant force is found by addition and subtraction. (Figure 2.8a) A total force of 180 N acting to the left and 120 N acting to the right will give a resultant force of $\overleftarrow{60 \text{ N}}$ acting to the left. The body is not in equilibrium and will require a force of $\overrightarrow{60 \text{ N}}$ acting to the right in the same line to bring the system to rest.

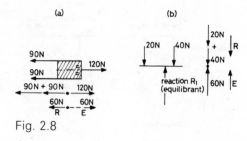

Fig. 2.8

In Figure 2.8b two forces of 20 N and 40 N act perpendicular on a rod. Neglecting the mass of the rod the resultant downward force is ↓ 60 N. The rod is not at rest, so a reaction R_1 (equilibrant) exerting ↑ 60 N vertically upwards is required to obtain equilibrium. The exact position of R_1 can be found by drawing but should be calculated (see worked example 4, page 59).

Method of finding the resultant of two non-parallel forces acting at a point

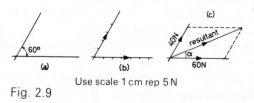

Use scale 1 cm rep 5 N

Fig. 2.9

Let the forces be 40 N and 60 N and the angle between them 60°. Set-off two light lines at an angle of 60° to each other (Figure 2.9a). Choose a suitable scale and mark off the magnitudes of the 40 N and 60 N forces along the lines (Figure 2.9b). Using set squares draw lines parallel to the vectors thus completing the Parallelogram of Forces (Figure 2.9c). The diagonal which is drawn from the intersection of the forces is measured and when multiplied by the scale factor the magnitude of the resultant force is obtained (87 N). The direction (sense) of the resultant force is shown. Using a protractor the angle α(24°) is determined.

When the resultant force is found the equilibrant (which can be a reaction or resistance) can thus be determined. The converse of this is true and by graphical means a single force can be broken up into two *component forces* having the same effect as the single force. Component forces will be dealt with in more detail in Chapter 3.

Method of drawing Three Coplanar Forces in equilibrium
This method is called *The Triangle of Forces* and is used to solve problems involving three coplanar forces whose lines of action meet at the same point.

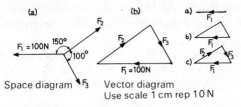

Space diagram   Vector diagram
              Use scale 1 cm rep 10 N

Fig. 2.10

The Space Diagram (Position diagram) (Figure 2.10a) is the name given to an accurate line drawing representing the action of forces in given problems. It is important that angles are accurately measured and that the line of action of each force is carefully drawn, as the accuracy of the results obtained from the vector diagram greatly depends on the accuracy of this diagram.

The Vector Diagram (Force diagram) (Figure 2.10b) is the name given to a diagram representing vector quantities. The magnitude and direction of forces can be determined from such a diagram.

Method of drawing the Vector diagram:

a) Draw the known vector (force) F_1 to a suitable scale. $F_1 = \overleftarrow{100\,\text{N}}$.

b) From each end of vector F_1 draw lines *parallel* to the lines of action of forces F_2 and F_3 *from the Space diagram*. This completes the vector triangle and indicates that the system is in equilibrium.

c) Insert arrowheads *in order* round the triangle.

d) The magnitude of forces F_2 and F_3 can be found by measuring the sides of the vector triangle and multiplying by the scale factor:

$$F_2 = (9\cdot55 \times 10)\,\text{N} \qquad F_3 = (5\cdot1 \times 10)\,\text{N}$$
$$= 95\cdot5\,\text{N} \qquad\qquad = 51\,\text{N}$$

Bow's Notation   This system consists of lettering each space between the forces in the *Space diagram using capital letters* (Figure 2.11a). Force F_1 (Figure 2.10a) becomes force AB, force F_2 becomes force BC and force F_3 becomes force CA. Notice that the letters must be taken in the same order, clockwise or anticlockwise.

The *Vector diagram* (Figure 2.11b) is lettered with corresponding *lower case letters*. As before, the triangle is constructed and the magnitude and direction of each force determined.

$$BC = (bc \times 10)\,N \qquad CA = (ca \times 10)\,N$$
$$= (9{\cdot}55 \times 10)\,N \qquad = (5{\cdot}1 \times 10)\,N$$
$$= 95{\cdot}5\,N \qquad\qquad = 51\,N$$

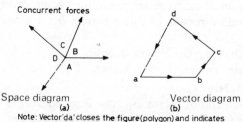

Space diagram
(a)

Vector diagram
(b)

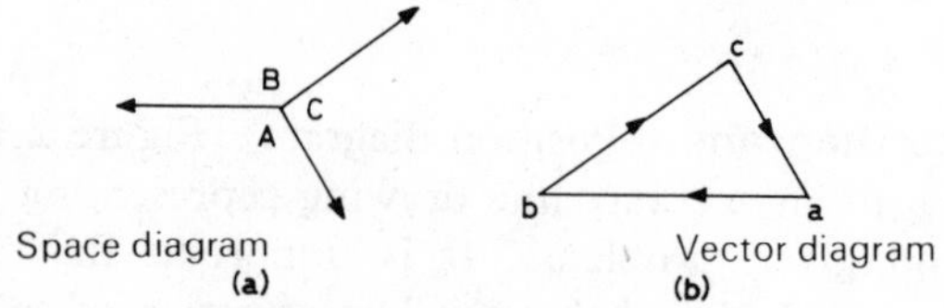

Fig. 2.11   Note: The closing vector of the triangle
is the equilibrant of the other two forces.

Method of drawing four or more coplanar forces in equilibrium. Polygon of Forces.

Concurrent forces

Space diagram
(a)

Vector diagram
(b)

Note: Vector 'da' closes the figure(polygon) and indicates
that the system is in equilibrium (no translation or rotation).
'DA' is the equilibrant of the system of forces.

Fig. 2.12

This method is similar to that used in drawing the Triangle of Forces. The Space diagram is drawn accurately and from it the Vector diagram is constructed. If the Vector diagram is a *closed polygon* the system is in equilibrium (Figure 2.12b).

In Figure 2.13b the Vector diagram is a different shape from that in Figure 2.12b although the figure is still a closed polygon. The forces can be transferred from the Space to Vector diagrams in any order, but care must be taken to see that the direction (sense) of one vector follows exactly that of the previous one.

Note that in Figure 2.13b difficulty would arise if use were made of Bow's Notation, as the forces are not taken in a clockwise or anticlockwise order.

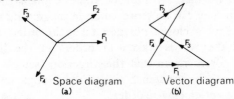

Fig. 2.13   Space diagram   Vector diagram
                                  (a)   (b)

In Figure 2.14a non-concurrent forces are shown acting on a beam. The Vector diagram (Figure 2.14b) is drawn. Note that forces F_1, F_2, F_3 and F_4 do not form a closed polygon. The closing force F_5 is the equilibrant and this force must be put into the system in some form (in the Space diagram) to obtain equilibrium. The polygon of forces can be used to determine the magnitude and direction of the equilibrant (or resultant) of a system of non-concurrent coplanar forces acting on a body, but the exact position or point of application of the force can best be determined by calculation (see Chapter 4: Moments).

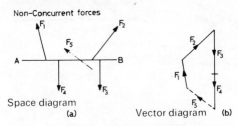

Fig. 2.14   Note: The closed figure(polygon) indicates no translation. The exact position of F_5 in the space diagram ensures no rotation and the system will then be in equilibrium.

Summary of Definitions

Parallelogram of Forces   If two forces acting at a point are represented in magnitude and direction by the adjacent sides of a parallelogram, then the diagonal through the point represents, in magnitude and direction, the *Resultant* of the two forces.

Triangle of Forces   If three forces are in equilibrium, they can be represented by the sides of a triangle, taken in order.

NOTE: If the body is in equilibrium, the three forces will be *coplanar* and their lines of actions *concurrent*.

Polygon of Forces   If four or more coplanar forces are in equilibrium, they can be represented in magnitude and direction by the sides of a closed polygon taken in order. It should be noted that the words 'taken in order' in the Triangle and Polygon of forces mean that the direction (sense) of the forces follow each other round each side of the triangle or polygon in a clockwise or anticlockwise order.

Worked Examples

1   Two forces of 20 N and 15 N act from a point O as shown. If the angle between them is 60°, find the magnitude and direction of the resultant force.

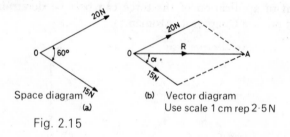

Space diagram
(a)

(b)   Vector diagram
Use scale 1 cm rep 2·5 N

Fig. 2.15

From Figure 2.15b:
R = The resultant force
R = (12·15 × 2·5) N   Angle α = 35° found by
  = 30·4 N   protractor.

The resultant force = 30·4 N acting at 35° from the 15 N force

2   The resultant of two unequal forces (inclined to each other at 80°) is 50 N and makes an angle of 30° with one of the other forces. Calculate the magnitude of the two forces.

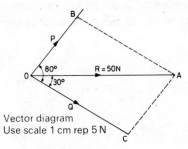

Fig. 2.16

Let the unknown forces be P and Q.
Draw the resultant force R to scale.
Through 'O' draw a line at 30° to represent the line of action of force Q.
Through 'O' draw a line at 80° from the line of action of force Q. This line represents the line of action of force P. From 'A' draw lines parallel to P and Q intersecting at 'C' and 'B' respectively. Measure OB and OC.

$$OB = P = (5 \cdot 1 \times 5)\,N \qquad OC = Q = (7 \cdot 8 \times 5)\,N$$
$$= 25 \cdot 5\,N \qquad\qquad = 39\,N$$

The forces are 25·5 N and 39 N

3   In a structure two members A and B are riveted to a bracket C as shown. If the force in each member is 80 kN find the resultant force in magnitude and direction acting on bracket C.

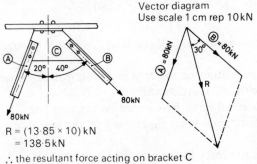

Vector diagram
Use scale 1 cm rep 10 kN

R = (13·85 × 10) kN
  = 138·5 kN

∴ the resultant force acting on bracket C
is 138·5 kN at 30° from member B

Fig. 2.17

4   Determine the horizontal and vertical components of the single
100 N force acting at 60° to the horizontal as shown.

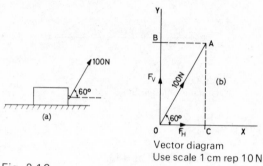

Vector diagram
Use scale 1 cm rep 10 N

Fig. 2.18

Let F_V and F_H be the vertical and horizontal components respectively.

(Figure 2.18b) Draw a horizontal line OX to represent the line of action of F_H.

Draw the vector OA at 60° to the horizontal to represent the 100 N force.

From 'O' draw OY perpendicular to OX to represent the line of action of F_V.

From 'A' draw AB and AC parallel to OX and OY respectively.

OB represents vector F_V

OC represents vector F_H

$$F_V = (8 \cdot 65 \times 10) \text{ N} \qquad F_H = (5 \times 10) \text{ N}$$
$$= 86 \cdot 5 \text{ N} \qquad\qquad = 50 \text{ N}$$

↑

The components are 86·5 N (vertical) and $\overrightarrow{50 \text{ N}}$ (horizontal)

5   A guy rope is attached to a tent peg as shown. If the tension in the
guy rope is 50 N, determine:

a) The vertical force F_V exerted on the tent peg.

b) The horizontal force F_H exerted on the tent peg.

Let F_I be the force in the guy rope.

Let F_V and F_H be the vertical and horizontal component forces
acting on the tent peg.

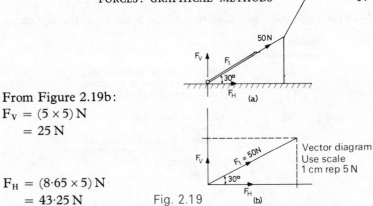

From Figure 2.19b:

$F_V = (5 \times 5)\,N$

$\quad = 25\,N$

$F_H = (8 \cdot 65 \times 5)\,N$

$\quad = 43 \cdot 25\,N$ Fig. 2.19

The vertical force is 25 N and the horizontal force is 43·25 N

6   A mass of 10 kg is raised by means of a rope passing over a friction-less pulley as shown. Determine:

   a) The magnitude of the force F_1.

   b) The magnitude and direction of the reaction R_1 at the pulley pin.

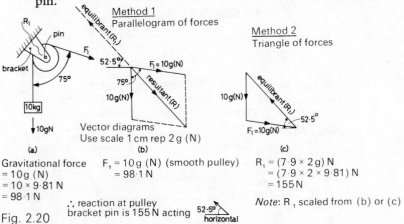

Fig. 2.20

7   A theatre arc light of mass 4 kg hangs from a ceiling. A chain supporting the light is pulled horizontally to the side by means of a string.

   Determine:

   a) The magnitude and nature of the forces in the chain ① and the string ②.

b) The magnitude and direction of the reaction at the ceiling.

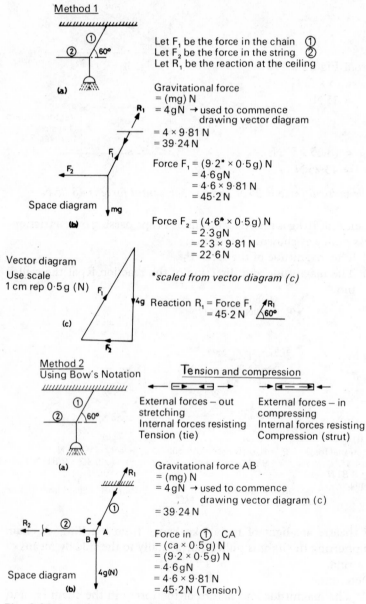

Method 1

Let F_1 be the force in the chain ①
Let F_2 be the force in the string ②
Let R_1 be the reaction at the ceiling

Gravitational force
= (mg) N
= 4 gN → used to commence
              drawing vector diagram
= 4 × 9·81 N
= 39·24 N

Force F_1 = (9·2* × 0·5g) N
       = 4·6gN
       = 4·6 × 9·81 N
       = 45·2 N

(a)

Space diagram

(b)

Force F_2 = (4·6* × 0·5g) N
       = 2·3gN
       = 2·3 × 9·81 N
       = 22·6 N

Vector diagram
Use scale
1 cm rep 0·5g (N)

scaled from vector diagram (c)

Reaction R_1 = Force F_1
         = 45·2 N

(c)

Method 2
Using Bow's Notation

Tension and compression

External forces – out
stretching
Internal forces resisting
Tension (tie)

External forces – in
compressing
Internal forces resisting
Compression (strut)

(a)

Gravitational force AB
= (mg) N
= 4 gN → used to commence
              drawing vector diagram (c)
= 39·24 N

Force in ① CA
= (ca × 0·5g) N
= (9·2 × 0·5g) N
= 4·6gN
= 4·6 × 9·81 N
= 45·2 N (Tension)

Space diagram

(b)

Fig. 2.21 (continued on next page)

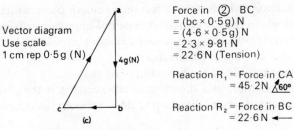

Vector diagram
Use scale
1 cm rep 0·5 g (N)

4g(N)

(c)

Force in ② BC
= (bc × 0·5 g) N
= (4·6 × 0·5 g) N
= 2·3 × 9·81 N
= 22·6 N (Tension)

Reaction R₁ = Force in CA
= 45·2 N ∠60°

Reaction R₂ = Force in BC
= 22·6 N ◄—

Fig. 2.21 (continued)

8   A garden roller of mass 60 kg is pulled up a wooden plank on to a lawn. If the roller is held on the plank in the position shown, determine:

a) The pull F required at the handle.

b) The reaction at the plank in magnitude and direction.

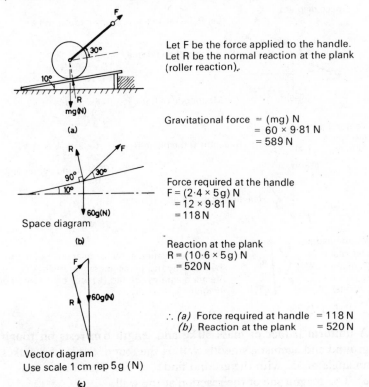

Let F be the force applied to the handle.
Let R be the normal reaction at the plank (roller reaction).

Gravitational force = (mg) N
= 60 × 9·81 N
= 589 N

Force required at the handle
F = (2·4 × 5g) N
= 12 × 9·81 N
= 118 N

Reaction at the plank
R = (10·6 × 5g) N
= 520 N

∴ *(a)* Force required at handle = 118 N
(b) Reaction at the plank = 520 N

Fig. 2.22

9   A box of mass 15 kg is held at rest on a smooth plane which is inclined at 30° to the horizontal. A rope holding the box is inclined at 20° to the plane, and passes over a smooth circular rod at X and is held at P as shown. Determine:

a) The magnitude of the force required at P.

b) The magnitude and direction of the reaction at the plane.

c) The magnitude and direction of the reaction at the circular rod.

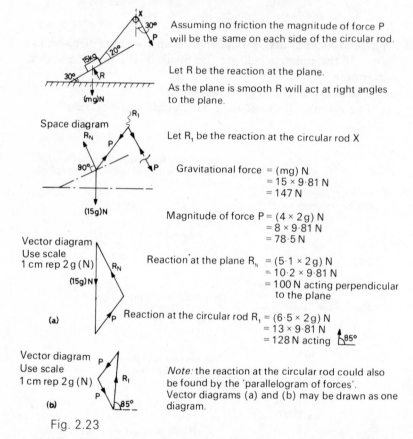

Assuming no friction the magnitude of force P will be the same on each side of the circular rod.

Let R be the reaction at the plane.

As the plane is smooth R will act at right angles to the plane.

Let R_1 be the reaction at the circular rod X

Gravitational force $= (mg)$ N
$= 15 \times 9{\cdot}81$ N
$= 147$ N

Magnitude of force $P = (4 \times 2g)$ N
$= 8 \times 9{\cdot}81$ N
$= 78{\cdot}5$ N

Reaction at the plane $R_N = (5{\cdot}1 \times 2g)$ N
$= 10{\cdot}2 \times 9{\cdot}81$ N
$= 100$ N acting perpendicular to the plane

Reaction at the circular rod $R_1 = (6{\cdot}5 \times 2g)$ N
$= 13 \times 9{\cdot}81$ N
$= 128$ N acting $85°$

Note: the reaction at the circular rod could also be found by the 'parallelogram of forces'.
Vector diagrams (a) and (b) may be drawn as one diagram.

Fig. 2.23

10 A uniform ladder of mass 30 kg and length 6 m rests on rough ground and against a smooth wall. If the foot of the ladder makes an angle of 55° with the ground find:

a) The magnitude of the reaction at the wall.

b) The magnitude and direction of the reaction at the ground.

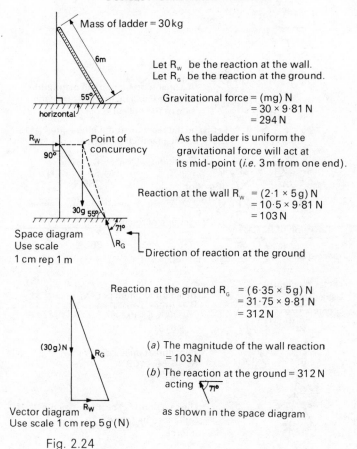

Mass of ladder = 30 kg

6m

55°
horizontal

Let R_w be the reaction at the wall.
Let R_G be the reaction at the ground.

Gravitational force = (mg) N
$$= 30 \times 9\cdot81 \text{ N}$$
$$= 294 \text{ N}$$

R_W
90°
Point of concurrency

As the ladder is uniform the gravitational force will act at its mid-point (i.e. 3 m from one end).

30g 55°
71°

Space diagram
Use scale
1 cm rep 1 m

R_G

Direction of reaction at the ground

Reaction at the wall $R_w = (2\cdot1 \times 5g)$ N
$$= 10\cdot5 \times 9\cdot81 \text{ N}$$
$$= 103 \text{ N}$$

Reaction at the ground $R_G = (6\cdot35 \times 5g)$ N
$$= 31\cdot75 \times 9\cdot81 \text{ N}$$
$$= 312 \text{ N}$$

(30g) N R_G

(a) The magnitude of the wall reaction
$$= 103 \text{ N}$$

(b) The reaction at the ground = 312 N
acting 71°

Vector diagram R_W
Use scale 1 cm rep 5g (N)

as shown in the space diagram

Fig. 2.24

11 A uniform square sectioned mast is hinged at X as shown. A force P applied at the end of a rope (which passes over a smooth pulley) holds the top of the mast (mass 80 kg and length 3 m) in equilibrium. Determine:
a) The magnitude of the force P.
b The reaction at the hinge X in magnitude and direction.
c) The magnitude and direction of the reaction at pulley pin Y.

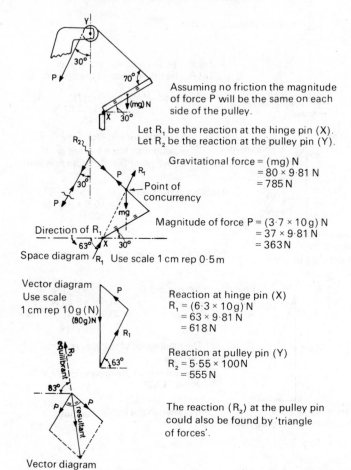

Assuming no friction the magnitude of force P will be the same on each side of the pulley.

Let R_1 be the reaction at the hinge pin (X).
Let R_2 be the reaction at the pulley pin (Y).

Gravitational force = (mg) N
$= 80 \times 9 \cdot 81$ N
$= 785$ N

Point of concurrency

Direction of R_1

Space diagram Use scale 1 cm rep 0·5 m

Magnitude of force P = $(3 \cdot 7 \times 10g)$ N
$= 37 \times 9 \cdot 81$ N
$= 363$ N

Vector diagram
Use scale
1 cm rep 10g (N)
(80g) N

Reaction at hinge pin (X)
$R_1 = (6 \cdot 3 \times 10g)$ N
$= 63 \times 9 \cdot 81$ N
$= 618$ N

Reaction at pulley pin (Y)
$R_2 = 5 \cdot 55 \times 100$ N
$= 555$ N

The reaction (R_2) at the pulley pin could also be found by 'triangle of forces'.

Vector diagram
Use scale 1 cm rep 100 N
Fig. 2.25

12 A circular post is held in position by four wire ropes acting in the same plane. If the ropes maintain the post in equilibrium find graphically the magnitude of the forces in ropes ① and ② .

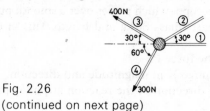

Fig. 2.26
(continued on next page)

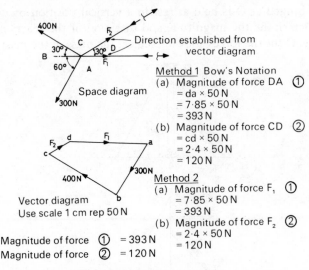

400N

C

F_2

Direction established from
vector diagram

30°
B
60°
30° D
F_1
A

Space diagram

300N

F_2 d F_1 a

c

400N

300N

Vector diagram b
Use scale 1 cm rep 50 N

Magnitude of force ① = 393 N
Magnitude of force ② = 120 N

Method 1 Bow's Notation
(a) Magnitude of force DA ①
$= da \times 50\,N$
$= 7.85 \times 50\,N$
$= 393\,N$
(b) Magnitude of force CD ②
$= cd \times 50\,N$
$= 2.4 \times 50\,N$
$= 120\,N$

Method 2
(a) Magnitude of force F_1 ①
$= 7.85 \times 50\,N$
$= 393\,N$
(b) Magnitude of force F_2 ②
$= 2.4 \times 50\,N$
$= 120\,N$

Fig. 2.26 (continued)

13 The steel plate structure shown is held in a vertical position by
four wire ropes acting in the same plane.
Determine:
a) The magnitude of the force in rope ③.
b) The gravitational force.
c) The mass of the plate.

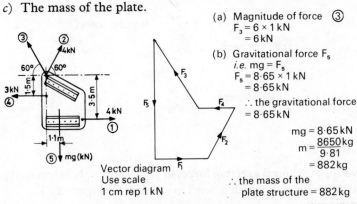

③
②
4kN

60° 60°

3kN
④

1.5m

3.5m

4kN
①

1.1m

⑤ mg(kN)

(a) Magnitude of force ③
$F_3 = 6 \times 1\,kN$
$= 6\,kN$

(b) Gravitational force F_5
i.e. $mg = F_5$
$F_5 = 8.65 \times 1\,kN$
$= 8.65\,kN$

∴ the gravitational force
$= 8.65\,kN$

$mg = 8.65\,kN$
$m = \dfrac{8650\,kg}{9.81}$
$= 882\,kg$

∴ the mass of the
plate structure = 882 kg

F_3

F_5
F_4

F_2

Vector diagram F_1
Use scale
1 cm rep 1 kN

Note: the forces in the vector diagram can be drawn in any
order, but must have the same direction (sense) round the
polygon. The use of Bow's Notation would cause some
difficulty in an example of this type where the known
forces are not adjacent.

Fig. 2.27

14 A mast hinged at O is held at rest in a vertical position by wire
    stays. Determine the magnitude and direction of the reaction at
    the hinge O for the forces shown. Neglect the mass of the mast.

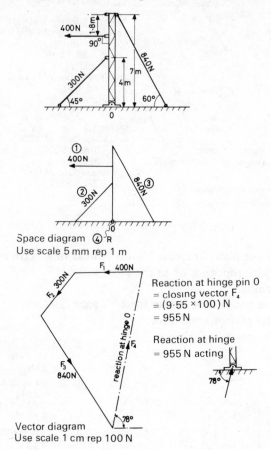

Space diagram
Use scale 5 mm rep 1 m

Reaction at hinge pin 0
= closing vector F₄
= (9·55 ×100) N
= 955 N

Reaction at hinge
= 955 N acting

Vector diagram
Use scale 1 cm rep 100 N

Fig. 2.28

Examples

1 Two ropes A and B exerting forces of 120 N and 80 N respectively
   pull on an eye bolt which is fixed to a wall at C (Figure 2.29).
   Determine:
   a) The magnitude of the force required by a single rope to replace
      the combined effect of both ropes.
   b) The magnitude and direction of the reaction at the wall.

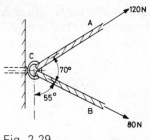

Fig. 2.29

2 Two unequal forces inclined to one another at 70° have a resultant force of 25 N which makes an angle of 30° with one of the forces. Find the magnitude of both forces.

3 A ship is towed at constant speed by pulls of two hawsers from tugs. If the tensions in the hawsers are 70 kN and 90 kN and the resistance to motion of the ship is 120 kN, find the angles made by the hawsers with the direction of motion.

4 A packing case is moved steadily along a horizontal floor by means of a rope inclined at an angle of 30° to the floor. If the tension in the rope is 400 N determine the effective horizontal and vertical forces acting on the case.

5 The figure shows a barge which is towed along a canal at constant speed. The axis of the barge is parallel to the tow path XX (7 m apart) (Figure 2.30). The towing rope which is 20 m long exerts a force of 160 N on the barge. Determine:

   a) The effective force in the direction of motion of the barge.

   b) The angle that the tow rope makes with the direction of motion of the barge.

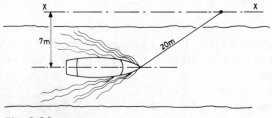

Fig. 2.30

6   Two timber posts A and B of equal length are inclined at 60° to the horizontal and fixed together at C (Figure 2.31). What force will each post exert to support a mass of 1 tonne suspended from C?

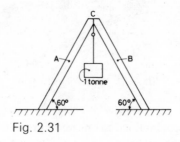

Fig. 2.31

7   A mass A of 20 kg is supported by a cord which passes over a frictionless pulley B and is fixed to a wall at C (Figure 2.32). Determine:
    a) The magnitude of the force in cord ①.
    b) The magnitude and direction of the reaction at the pulley pin D.

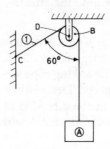

Fig. 2.32

8   In a structure three members A, B and C are riveted to a bracket plate D (Figure 2.33). The tension in member A is 40 kN and the resultant of the tensions in members A and B is to act along XX. Determine:
    a) The tension in member B.
    b) The magnitude and direction of the force in member C to maintain equilibrium.

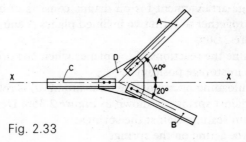

Fig. 2.33

9 Each figure shows a mass 'A' supported by ropes in various positions. Determine the tension in each rope for the positions shown (Figure 2.34) if mass 'A' is 55 kg.

Length of rope ① = 2 m.
Length of rope ② = 3 m

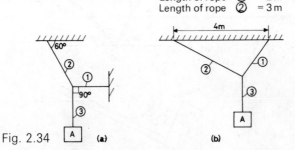

Fig. 2.34 (a) (b)

10 A conveyor system consisting of metal rollers is used to lower packages from one level to another. The conveyor is inclined at 20° to the horizontal. Two packages each of 30 kg are tied together and held on the conveyor by means of a rope (Figure 2.35). Determine the force in the rope to hold the packages if:
a) The rope is parallel to the conveyor.
b) The rope is at an angle of 20° above the conveyor.

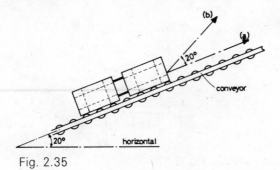

Fig. 2.35

11 a) A storage arrangement for oil drums consists of metal plates
        welded together to form two inclined planes A and B as shown
        in Figure 2.36a.
        Determine the reactions of the planes when one drum of mass
        60 kg is in storage position.

    b) In an unloading operation the oil drum in (a) is rolled down a
        ramp against springs as shown in Figure 2.36b. Determine for
        one drum resting against the springs:

    i) The force acting on the springs.

    ii) The reaction at the ramp.

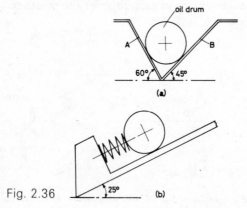

(a)

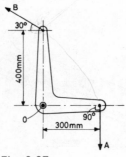

Fig. 2.36   (b)

12 A right angled bell crank lever pivoted at O is in equilibrium due
    to the action of two forces A and B (Figure 2.37). If force A is 60 N
    determine:

    a) The magnitude of force B.

    b) The magnitude and direction of the reaction at the pivot pin O.

Fig. 2.37

13 A machine shaft is shown (Figure 2.38) supported by two ropes
   A and B. If the mass of the shaft is 100 kg determine:
   a) The tension in each rope.
   b) The position of the centre of gravity of the shaft (from the left
      hand end).

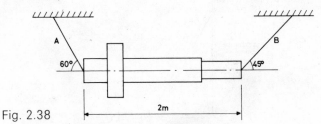

Fig. 2.38

14 A uniform rod BC of mass 25 kg and length 3 m is hinged at C and
   held in the position shown by a cord AB which is connected to a
   wall at A (Figure 2.39).
   Determine:
   a) The tension in cord AB.
   b) The magnitude and direction of the reaction at A.
   c) The magnitude and direction of the reaction at the hinge pin C.

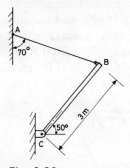

Fig. 2.39

15 During the construction of a ship a uniform ladder is placed against
   a steel deckhouse casing (assumed smooth) with its lower end
   resting against an angle bar which is fixed to the deck. A tradesman
   of mass 75 kg (carrying two tins of paint, each of mass 2·5 kg)
   stands at the middle of the ladder.
   Neglecting the mass of the ladder, and for the position shown
   (Figure 2.40) determine:
   a) The length of the ladder.

b) The magnitude and direction of the reaction at the deck.
c) The magnitude of the reaction at the deckhouse casing.

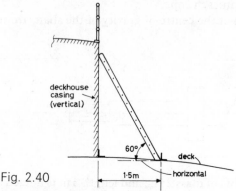

Fig. 2.40

16 A uniform ladder 5 m long rests against a smooth vertical wall with its lower end inclined at an angle of 50° to rough horizontal ground. If the reaction at the wall is 60 N determine:
a) The mass of the ladder.
b) The magnitude and direction of the reaction at the ground.

17 An overhead platform for inspecting street lights is shown. The platform has a mass of 125 kg, and a man of mass 75 kg stands at its centre. For the position shown (Figure 2.41) find:
a) The force on the hydraulic piston A.
b) The magnitude and direction of the reaction at the pivot pin O. Neglect the mass of the arm.

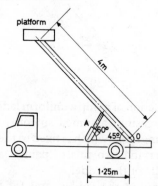

Fig. 2.41

18 Two pieces of wood acting as props hold a uniform desk lid of mass 2 kg in a raised position shown (Figure 2.42). If each prop

is in line with each hinge, find:
a) The magnitude of the force in each prop.
b) The magnitude and direction of the reaction at each hinge pin O.

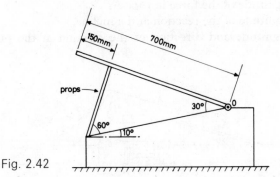

Fig. 2.42

19 The lid of a refuse bin is held in a raised position by means of a
cord and pulley arrangement (Figure 2.43). Assuming the pulley
to be frictionless and lid uniform determine:
a) The mass of the lid.
b) The reaction at the hing pin O in magnitude and direction.

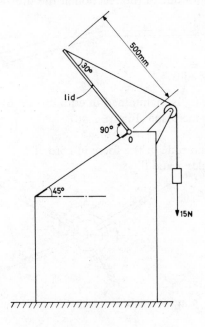

Fig. 2.43

20 A hauling arrangement for loading trucks is shown. If the mass of
    each crate being loaded is 300 kg and the rope is parallel to the
    smooth incline, determine for the position shown (Figure 2.44):
    a) The magnitude of the force in rope A.
    b) The magnitude of the reaction at the incline.
    c) The magnitude and direction of the reaction at the pulley
       pin O.

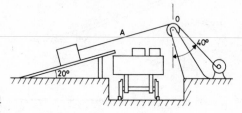

Fig. 2.44

21 A rowing boat is hauled up a smooth slipway at constant speed.
    The rope which passes over a smooth roller at A exerts a force of
    540 N (Figure 2.45).
    Determine:
    a) The mass of the boat.
    b) The magnitude of the reaction at the slipway.

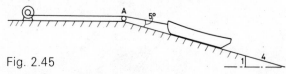

Fig. 2.45

22 A smooth ring is maintained in equilibrium by four cords which
    act in the same plane (Figure 2.46).
    Find:
    a) The magnitude of the force in cord ①.
    b) The angles α and θ.

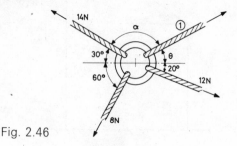

Fig. 2.46

23 A metal plate of mass 80 kg is supported by a number of wire
    ropes as shown (Figure 2.47). If the ropes are coplanar and assum-
    ing no friction at the pulley determine the force in each of the
    ropes A to F.

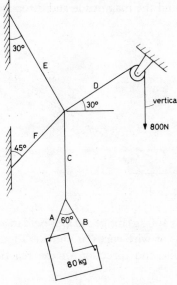

Fig. 2.47

24 The jib crane shown carries a mass of 300 kg. Assuming no friction
    at the pulley find for the position shown (Figure 2.48):
    a) The magnitude of the forces in the Jib AB and the Tie AC.
    b) The magnitude and direction of the reactions at B and C.

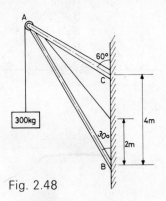

Fig. 2.48

25   A uniform beam OB 6 m long is at rest in the inclined position shown (Figure 2.49). The beam is hinged at O and rests on a roller at P. If the gravitational force acting on the beam is 4 kN, the vertical downward force at B 6 kN and the reaction at the roller P 10·4 kN, find the magnitude and direction of the reaction at the hinge pin O.

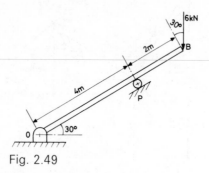

Fig. 2.49

26   A metal mast (mass 100 kg) hinged at O is held in a vertical position by an arrangement of wire ropes as shown (Figure 2.50). Determine the magnitude and direction of the reaction at the hinge pin O.

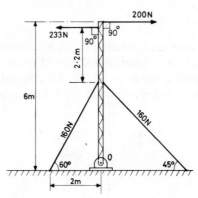

Fig. 2.50

3
Forces: Analytical Methods

In graphical treatment of force systems accuracy greatly depends on the exactness of measurement and the correct manipulation of drawing instruments. Although this method is quite satisfactory for most practical problems more accurate results can be obtained by calculation.

In the previous chapter examples were considered where two forces acting at a point could be replaced by a single force which would have the same effect. The converse of this is true: a single force can be converted into two forces which together will have the same effect as the original force. The two forces are called *components* of the single force. Figure 3.1 shows that a single force can be *resolved* into an unlimited number of pairs of component forces, the single force R being the diagonal of a number of parallelograms.

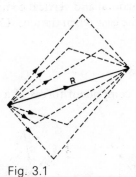

Fig. 3.1

Rectangular Components   A method often used in engineering calculations is that of converting a single force into two component forces which are at right angles to each other (Figure 3.2).

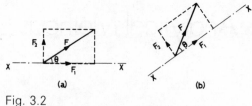

Fig. 3.2

To find the rectangular component forces F_1 and F_2 of a given single force F:

Consider the single force F at angle θ to XX.

Using the single force F as the diagonal construct a rectangle on XX.

By trigonometry:

$$\frac{F_1}{F} = \cos \theta$$

$$F_1 = F \cos \theta$$

$$\text{and } \frac{F_2}{F} = \sin \theta$$

$$F_2 = F \sin \theta$$

$F \cos \theta$ is the component of force F parallel to XX

$F \sin \theta$ is the component of force F perpendicular to XX

NOTE: In Figure 3.2a the component forces F_1 and F_2 are often referred to as the horizontal and vertical components of the single force F. This is only the case when the axis XX is horizontal.

Worked Examples

1 Find the rectangular component forces of a single force of 10 N acting at 60° to XX.

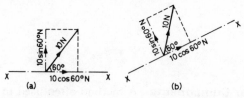

Fig. 3.3

Resolving the 10 N force parallel to XX
$$= 10 \cos 60° \, \text{N}$$
$$= 10 \times 0.5 \, \text{N}$$
$$= 5 \, \text{N}$$

Resolving the 10 N force perpendicular to XX
$$= 10 \sin 60° \, \text{N}$$
$$= 10 \times 0.866 \, \text{N}$$
$$= 8.66 \, \text{N}$$

The rectangular component forces are 5 N and 8·66 N

2   Calculate the sum of the rectangular components of each force shown and determine the magnitude and direction of the resultant force of the system of forces.

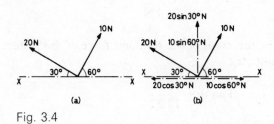

(a)   (b)

Fig. 3.4

The positive direction of the component forces can usually be determined by observation. If a negative sign appears in an answer the direction has probably been wrongly assumed and should be reversed.

From Figure 3.4b:
Resolving parallel to XX (horizontal)

$$\Sigma_c \overset{+}{\overleftarrow{XX}} = (20 \cos 30° - 10 \cos 60°) \, \text{N}$$
$$= (20 \times 0.866 - 10 \times 0.5) \, \text{N}$$
$$= 17.32 \, \text{N} - 5 \, \text{N}$$
$$= 12.32 \, \text{N} \ (\text{direction correctly assumed})$$

Resolving perpendicular to XX (vertical)

$$+\uparrow \ \Sigma_c \perp XX = (10 \sin 60° + 20 \sin 30°) \, \text{N}$$
$$= (10 \times 0.866 + 20 \times 0.5) \, \text{N}$$
$$= 8.66 \, \text{N} + 10 \, \text{N}$$
$$= 18.66 \, \text{N} \ (\text{direction correctly assumed})$$

To find the resultant force of the system:

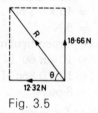

Fig. 3.5

By theorem of Pythagoras

$$R = \sqrt{18 \cdot 66^2 + 12 \cdot 32^2} \, N$$
$$= \sqrt{500} \, N$$
$$= 22 \cdot 4 \, N$$

To find the direction of the resultant force of the system:
By trigonometry

$$\text{Tan } \theta = \frac{18 \cdot 66}{12 \cdot 32}$$

$$\text{Tan } \theta = 1 \cdot 512$$
$$\theta = 56 \cdot 5°$$

Alternative method of finding the magnitude of resultant force R
having found $\theta = 56 \cdot 5°$:

$$\text{Cos } 56 \cdot 5° = \frac{12 \cdot 32}{R}$$

$$R = \frac{12 \cdot 32}{\text{Cos } 56 \cdot 5°} \, N$$

$$R = \frac{12 \cdot 32}{0 \cdot 552} \, N$$

$$R = 22 \cdot 4 \, N$$

The resultant force of the system of forces = 22·4 N acting 56·5°⟋

NOTE: If the above system is to be in equilibrium a force E of
22·4 N acting ⟍56·5° would have to be placed into the system as
shown in Figure 3.6.

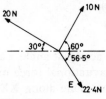

Fig. 3.6

NOTE: for equilibrium

a) the algebraic sum of the component forces along XX must be zero. $\Sigma_C XX = 0$ or $\Sigma_C \overleftarrow{XX} = \Sigma_C \overrightarrow{XX}$.

b) the algebraic sum of the component forces perpendicular to XX must be zero. $\Sigma_C \perp XX = 0$ or $\Sigma_C \perp XX\downarrow = \Sigma_C \perp XX\uparrow$.

c) the algebraic sum of the moments of the forces about any point in the plane must be zero. $\Sigma M = 0$ or (Σ A.C.M. = Σ C.M.).

(Σ is the Greek letter 'sigma' and means 'the sum of'.)

The above conditions for equilibrium are illustrated by worked examples in this chapter.

The Inclined Plane In this chapter the plane is assumed smooth (see also Chapter 2, worked example 9).

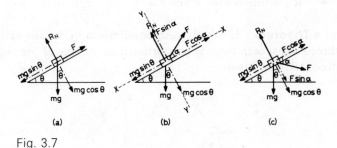

(a)   (b)   (c)

Fig. 3.7

In each case the gravitational force (mg) is resolved into rectangular components. Resolving mg parallel to the plane = mg sin θ and resolving mg perpendicular to the plane = mg cos θ. Figure 3.7a: Two equations are obtained from which F and R_N can be obtained:

$$F = mg \sin \theta \quad \ldots \ldots \ldots \quad ①$$
$$R_N = mg \cos \theta \quad \ldots \ldots \ldots \quad ②$$

Figure 3.7b: Force F is acting at angle α above the plane.
The force F and the gravitational force mg are resolved parallel and perpendicular to the plane along XX and YY respectively. The two equations are:

$$F \cos \alpha = mg \sin \theta$$

$$F = \frac{mg \sin \theta}{\cos \alpha} \quad \ldots \ldots \ldots \quad ①$$

$$R_N + F \sin \alpha = mg \cos \theta$$
$$R_N = mg \cos \theta - F \sin \alpha \quad \ldots \quad ②$$

Figure 3.7c: Force F is acting at angle α below the plane. The equations are:

$$F \cos \alpha = mg \sin \theta$$

$$F = \frac{mg \sin \theta}{\cos \alpha} \quad \ldots \ldots \ldots \quad ①$$

$$R_N = mg \cos \theta + F \sin \alpha \quad \ldots \quad ②$$

If force F is horizontal $\alpha = \theta$
$$F = mg \tan \theta \quad \ldots \ldots \ldots \quad ①$$
$$R_N = mg \cos \theta + F \sin \theta \quad \ldots \quad ②$$

Lami's Theorem   If a body is in equilibrium under the action of three forces, each force is proportional to the sine of the angle between the other two.

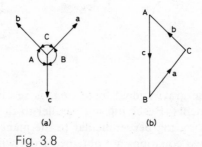

(a)   (b)

Fig. 3.8

Figure 3.8a:

$$\frac{a}{\sin A} = \frac{b}{\sin B} = \frac{c}{\sin C}$$

This is an extension of the 'sine rule' as can be seen from Figure 3.8b.

NOTE: when using Lami's theorem, if angle $\theta > 90°$ and $< 180°$ then $\sin \theta = \sin (180 - \theta)$

3 Three forces acting at a point are in equilibrium. If two of the forces are 20 N and 30 N find the magnitude and direction of the third force.

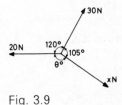

Fig. 3.9

Let the unknown force be 'x' N
Let the unknown angle be $\theta°$
Angle $\theta = [360° - (120° + 105°)]$ geometry.
    $\theta = 360° - 225°$
    $\theta = 135°$
By Lami's theorem

$$\underset{\textcircled{1}}{\frac{20}{\sin 105°}} = \underset{\textcircled{2}}{\frac{30}{\sin 135°}} = \underset{\textcircled{3}}{\frac{'x'}{\sin 120°}}$$

Using ① and ③

$$\frac{20}{\sin (180° - 105°)} = \frac{x}{\sin (180° - 120°)}$$

$$x = \frac{20 \sin 60°}{\sin 75°}$$

$$x = 18$$

The magnitude of the third force is 18 N and angle $\theta = 135°$

4   Solution to worked example, question 1, Chapter 2.

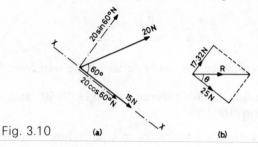

Fig. 3.10   **(a)**   **(b)**

Resolving along XX
$$+\searrow \Sigma_C XX = (15 + 20 \cos 60°)\,N$$
$$= 15\,N + 10\,N$$
$$= 25\,N$$
Resolving perpendicular to XX
$$+\nearrow \Sigma_C \perp XX = 20 \sin 60°\,N$$
$$= 17·32\,N$$

$$R = \sqrt{17·32^2 + 25^2}\,N$$
$$= 30·4\,N$$

$$\text{Tan}\,\theta = \frac{17·32}{25}$$

$$\text{Tan}\,\theta = 0·694$$
$$\theta = 34·8°$$

The resultant force = 30·4 N acting 34·8° from the 15 N force.

5   Solution to worked example, question 4, Chapter 2.

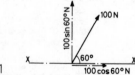

Fig. 3.11

Resolving along XX

$$\overset{+}{\overrightarrow{\Sigma_C XX}} = 100 \cos 60°\,N$$
$$= 100 \times 0·5\,N$$
$$= 50\,N$$

Resolving perpendicular to XX

$$+ \uparrow \Sigma_C \perp XX = 100 \sin 60° \text{ N}$$
$$= 100 \times 0.866 \text{ N}$$
$$= 86.6 \text{ N}$$

The horizontal component force = 50 N
The vertical component force = 86.6 N

6   Solution to worked example, question 7, Chapter 2.

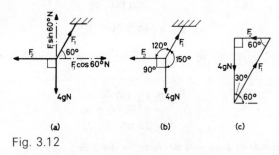

(a)   (b)   (c)

Fig. 3.12

Figure 3.12a, method 1 : resolution of forces.
Let F_1 be the force in the chain.
Let F_2 be the force in the string.

Resolving vertically
$$F_1 \sin 60° = 4 \text{ g N}$$

$$F_1 = \frac{4 \times 9.81 \text{ N}}{0.866}$$

$$F_1 = 45.3 \text{ N}$$

Resolving horizontally
$$F_2 = F_1 \cos 60°$$
$$F_2 = 45.3 \times 0.5 \text{ N}$$
$$F_2 = 22.65 \text{ N}$$

a) *Magnitude of the force in chain ① = 45.3 N*
b) *Magnitude of the force in string ② = 22.65 N*

Figure 3.12b, method 2 : Lami's theorem.
Let F_1 be the force in the chain.
Let F_2 be the force in the string.

$$\frac{F_1}{\sin 90°} = \frac{F_2}{\sin 150°} = \frac{4 \text{ g N}}{\sin 120°}$$

NOTE:

$\sin 90° = 1$ | $\sin 150° = \sin(180° - 150°)$ | $\sin 120° = \sin(180° - 120°)$
$= \sin 30°$ | $= \sin 60°$

$$\overset{①}{\qquad} \overset{②}{\qquad} \overset{③}{\qquad}$$

$$F_1 = \frac{F_2}{\sin 30°} = \frac{4g\,N}{\sin 60°}$$

① & ③ $\quad F_1 = \dfrac{4 \times 9\cdot81\ N}{\sin 60°}$

$$F_1 = 45\cdot3\ N$$

① & ② $\quad F_1 = \dfrac{F_2}{\sin 30°}$

$$\begin{aligned} F_2 &= F_1 \sin 30° \\ &= 45\cdot3 \times 0\cdot5\ N \\ &= 22\cdot65\ N \end{aligned}$$

a) *Magnitude of the force in chain = 45·3 N*
b) *Magnitude of the force in the string = 22·65 N*

Figure 3.12c, method 3 : by drawing the triangle of forces (not to scale) and calculating the forces F_1 and F_2 (*i.e.* the sides of the vector triangle)

$$\mathrm{Sin}\ 60° = \frac{4g\,N}{F_1}$$

$$F_1 = \frac{4g}{\sin 60°}\ N$$

$$= \frac{4 \times 9\cdot81}{0\cdot866}\ N$$

$$= 45\cdot3\ N$$

$$\mathrm{Cos}\ 60° = \frac{F_2}{F_1}$$

$$\begin{aligned} F_2 &= F_1 \times \cos 60° \\ &= 45\cdot3 \times 0\cdot5\ N \\ &= 22\cdot65\ N \end{aligned}$$

a) *Magnitude of the force in the chain = 45·3 N*
b) *Magnitude of the force in the string = 22·65 N*

7   Solution to worked example, question 8, Chapter 2.

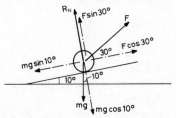

Fig. 3.13

Let F be the force applied to the handle.
Let R_N be the normal reaction at the plane (roller reaction).
Resolving parallel to the plane

$$F \cos 30° = mg \sin 10°$$
$$0·866 \, F = 60 \times 9·81 \times 0·174 \, N$$
$$F = 118 \, N$$

Resolving perpendicular to the plane

$$R_N + F \sin 30° = mg \cos 10°$$
$$R_N = (60 \times 9·81 \times 0·985) \, N - (118 \times 0·5) \, N$$
$$R_N = 579 \, N - 59 \, N$$
$$R_N = 520 \, N$$

a) *The pull at the handle = 118 N*
b) *The normal reaction at the plane = 520 N*

8   Solution to worked example, question 9, Chapter 2.

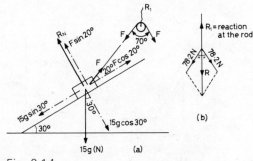

Fig. 3.14

Let F be the force in the rope.
Let R_N be the normal reaction at the plane.
Let R_1 be the reaction at the circular rod.

Resolving the gravitational force (mg) and F parallel to the plane
$$F \cos 20° = 15\,g \sin 30°$$

$$F = \frac{15 \times 9\cdot81 \times 0\cdot5 \text{ N}}{0\cdot94}$$

$$F = 78\cdot2 \text{ N}$$

Resolving mg and F perpendicular to the plane
$$R_N + F \sin 20° = 15\,g \cos 30°$$
$$R_N = (15 \times 9\cdot81 \times 0\cdot866)\,N - (78\cdot2 \times 0\cdot342)\,N$$
$$R_N = 127\,N - 27\,N$$
$$R_N = 100\,N$$

To find the reaction R_1 at the circular rod.

From Figure 3.14b, the magnitude of force F will be the same on each side of the rod.

The line of action of R_1 will bisect the angle between the two equal forces.

$$R_1 = 2F \cos \frac{\theta}{2} \text{ (where } \theta \text{ is the angle between the forces)}$$
$$R_1 = 2 \times 78\cdot2 \times \cos 35° \text{ N}$$
$$R_1 = 128 \text{ N}$$

a) *The force in the rope = 78·2 N*
b) *The normal reaction at the plane = 100 N*
c) *The reaction at the circular rod = 128 N*

9   Solution to worked example, question 10, Chapter 2.

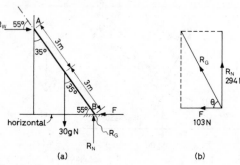

Fig. 3.15

Method 1 :

Let R_W be the reaction at the wall.

Let R_G be the reaction at the ground.

Let R_N be the vertical component of R_G.

Let F be the horizontal component of R_G.

Take moments about B (this eliminates two forces F and R_N).

For equilibrium

$$\Sigma \text{ C.M.} = \Sigma \text{ A.C.M.}$$
$$R_W \sin 55 \times 6 = 30 g \sin 35 \times 3$$

$$R_W = \frac{30 \times 9 \cdot 81 \times 0 \cdot 574 \times 3}{0 \cdot 82 \times 6} \text{ N}$$

$$R_W = 103 \text{ N}$$
$$\left. \begin{array}{l} R_W = F = 103 \text{ N} \\ R_N = 30 g \text{ N} = 294 \text{ N} \end{array} \right\} \text{ Action and reaction}$$

To find the reaction R_G at the ground: consider Figure 3.15b

$$R_G = \sqrt{294^2 + 103^2} \text{ N}$$
$$R_G = 312 \text{ N}$$

$$\text{Tan } \theta = \frac{294}{103}$$

$$\text{Tan } \theta = 2 \cdot 85$$
$$\theta = 71°$$

The reaction at the wall = 103 N

The reaction at the ground = 312 N acting $\overset{71°}{\nearrow}$

Method 2 : using Lami's theorem

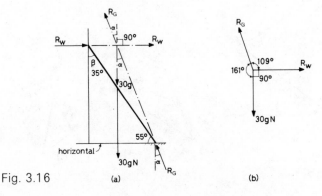

Fig. 3.16   (a)   (b)

From Figure 3.1 a: as the ladder is uniform

$$\text{Tan } \alpha = 0{\cdot}5 \tan \beta$$
$$\text{Tan } \alpha = 0{\cdot}5 \times \tan 35°$$
$$\text{Tan } \alpha = 0{\cdot}5 \times 0{\cdot}7$$
$$\text{Tan } \alpha = 0{\cdot}35$$
$$\alpha = 19°$$

Figure 3.16b: using Lami's theorem

$$\frac{R_G}{\sin 90°} = \frac{30\,g\,N}{\sin 109°}$$

$$R_G = \frac{30 \times 9{\cdot}81}{\sin 71°}\,N$$

$$R_G = 312\,N$$

$$\frac{R_W}{\sin 161°} = \frac{30\,g\,N}{\sin 109°}$$

$$R_W = \frac{30 \times 9{\cdot}81 \times \sin 19°}{\sin 71°}\,N$$

$$R_W = 103\,N$$

a) *The reaction at the wall = 103 N*
b) *The reaction at the ground = 312 N* 71°

10   Solution to worked example, question 11, Chapter 2.

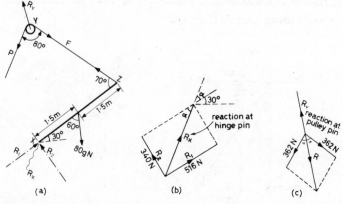

Fig. 3.17

Consider Figure 3.17a.

Let F be the force in the rope. Then $P = F$ (smooth pulley).

Let R_X be the reaction at the hinge pin.

Let R_1 and R_2 be the rectangular components of R_X.

Let R_Y be the reaction at the pulley pin Y.

Take moments about the hinge (X). This eliminates R_1 and R_2 and also the components of the gravitational force (mg) and force F along the top of the mast XZ.

For equilibrium

$$\Sigma \text{ C.M.} = \Sigma \text{ A.C.M.}$$
$$80 \text{ g sin } 60° \times 1·5 = F \sin 70° \times 3$$
$$F = \frac{80 \times 9·81 \times 0·866 \times 1·5}{0·94 \times 3} \text{ N}$$
$$F = 362 \text{ N}$$

To find the reaction at the hinge pin X.

Resolving along XZ

$+ \swarrow$ Assumed positive

$$F \cos 70° + 80 \text{ g} \cos 60° - R_1 = 0$$
$$R_1 = (362 \times 0·342) \text{ N} + (80 \times 9·81 \times 0·5) \text{ N}$$
$$R_1 = 124 \text{ N} + 392 \text{ N}$$
$$R_1 = 516 \text{ N}$$

Resolving perpendicular to XZ

$+ \searrow$ Assumed positive

$$80 \text{ g} \sin 60° - R_2 - F \sin 70° = 0$$
$$R_2 = (80 \times 9·81 \times 0·866) \text{ N} - (362 \times 0·94) \text{ N}$$
$$R_2 = 680 \text{ N} - 340 \text{ N}$$
$$R_2 = 340 \text{ N}$$

To find the reaction at the hinge pin X. Consider Figure 3.17b

$$R_X = \sqrt{516^2 + 340^2} \text{ N}$$
$$= 618 \text{ N}$$
$$\text{Tan } \alpha = \frac{340}{516}$$
$$\text{Tan } \alpha = 0·659$$
$$\alpha = 33·4°$$

To find the reaction at the pulley pin Y. Consider Figure 3.17c

$F = P$ (smooth pulley).

The line of action of reaction R_Y will bisect the angle between the equal forces.

$$R_Y = 2\,F \cos \frac{\theta}{2}\,N$$
$$= 2 \times 362 \times \cos\ 40°\,N$$
$$= 555\,N$$

a) *The force in the rope = 362 N*
b) *The reaction at the hinge pin = 618 N*
c) *The reaction at the pulley pin = 555 N*

11 Solution to worked example, question 12, Chapter 2.

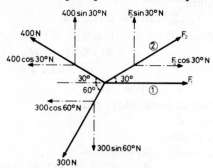

Fig. 3.18

Let F_1 and F_2 be the forces in ropes ① and ②.

Resolving the forces horizontally

$$(400 \cos 30° + 300 \cos 60°)\,N = F_1 + F_2 \cos 30°$$
$$(400 \times 0\cdot866 + 300 \times 0\cdot5)\,N = F_1 + 0\cdot866\,F_2$$
$$347\,N + 150\,N = F_1 + 0\cdot866\,F_2$$
$$F_1 + 0\cdot866\,F_2 = 497\,N$$

Resolving the forces vertically

$$F_2 \sin 30° + 400 \sin 30°\,N = 300 \sin 60°\,N$$
$$F_2 \times 0\cdot5 = 300 \times 0\cdot866\,N - 400 \times 0\cdot5\,N$$
$$F_2 = \frac{260 - 200}{0\cdot5}\,N$$
$$F_2 = 120\,N$$

By substitution $\quad F_1 = 497\,N - 0\cdot866\,F_2$
$$= 497\,N - 0\cdot866 \times 120\,N$$
$$= 497\,N - 104\,N$$
$$= 393\,N$$

a) *The magnitude of the force in rope* ① *= 393 N*
b) *The magnitude of the force in rope* ② *= 120 N*

12 Solution to worked example, question 13, Chapter 2.

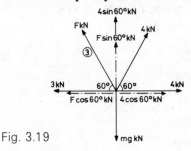

Fig. 3.19

Let F be the force in rope ③.
Resolving horizontally

$$3 \text{ kN} + F \cos 60° = 4 \text{ kN} + 4 \cos 60° \text{ kN}$$
$$0.5 \text{ F} = (4 + 4 \times 0.5) \text{ kN} - 3 \text{ kN}$$
$$F = 6 \text{ kN}$$

Resolving vertically

$$mg = (F \sin 60° + 4 \sin 60°) \text{ kN}$$
$$mg = (6 \times 0.866 + 4 \times 0.866) \text{ kN}$$
$$mg = 5.2 \text{ kN} + 3.46 \text{ kN}$$
$$= 8.66 \text{ kN (gravitational force)}$$
$$m = \frac{8\,660}{9.81} \text{ kg}$$
$$m = 883 \text{ kg}$$

a) *The magnitude of the force in the rope* = *6 000 N*
b) *The gravitational force* = *8 660 N*
c) *The mass of the plate* = *883 kg*

13 Solution to worked example, question 14, Chapter 2.

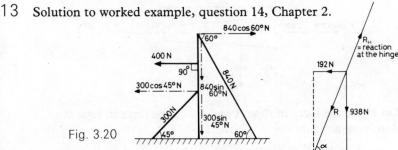

Fig. 3.20

(a)   (b)

Resolving horizontally

$\overset{+}{\leftarrow}$ Assumed positive

$$\Sigma_C H = 400\,N + 300\cos 45°\,N - 840\cos 60°\,N$$
$$= 400\,N + 212\,N - 420\,N$$
$$= 192\,N$$

Resolving vertically
$+\downarrow$ Assumed positive
$$= 300\sin 45°\,N + 840\sin 60°\,N$$
$$= 212\,N + 726\,N$$
$$= 938\,N$$

Consider Figure 3.20b
$$R = \sqrt{938^2 + 192^2}\,N$$
$$= 956\,N$$
$$\text{Tan }\alpha = \frac{938}{192}$$
$$\text{Tan }\alpha = 4\cdot 89$$
$$\alpha = 78°$$

The reaction at the hinge = 956 N 78°

14 A machine shaft is supported by two ropes A and B in the position
shown. Calculate:
a) The tension in each rope
b) The position of the centre of gravity of the shaft (from left
hand end).

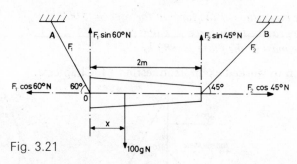

Fig. 3.21

Let F_1 be the force in rope A and F_2 be the force in rope B.
Resolving horizontally
$$F_1 \cos 60° = F_2 \cos 45°$$
$$F_1 = \frac{F_2 \cos 45°}{\cos 60°}$$

$$F_1 = \frac{0.707\,F_2}{0.5}$$
$$= 1.414\,F_2$$

Resolving vertically

$$F_1 \sin 60° + F_2 \sin 45° = mg\,N$$
$$1.414\,F_2 \times 0.866 + F_2 \times 0.707 = 100 \times 9.81\,N$$
$$1.223\,F_2 + 0.707\,F_2 = 981\,N$$
$$1.93\,F_2 = 981\,N$$
$$F_2 = 508\,N$$

By substitution

$$F_1 = 1.414\,F_2$$
$$= 1.414 \times 508\,N$$
$$= 718\,N$$

To find the position of the centre of gravity from the left hand end.

Take moments about O

For equilibrium

$$\Sigma\ \text{C.M.} = \Sigma\ \text{A.C.M.}$$
$$mg \times \text{'X'} = F \sin 45° \times 2$$
$$100 \times 9.81 \times X = 508 \times 0.707 \times 2\,m$$
$$X = \frac{718}{981}\,m$$
$$X = 0.732\,m$$
$$X = 732\,mm$$

a) *The tension in rope A = 718 N*
   The tension in rope B = 508 N

b) *The C. of G. of the shaft is 732 mm from the left hand end.*

NOTE: The tensions in the ropes could be found using Lami's theorem instead of resolution of forces, but the Principle of Moments must be applied to find the position of the centre of gravity.

Examples

Examples 1 to 26 of Chapter 2, already solved graphically, can now be solved analytically and results compared.

4
Moments

The *Moment* of a force is its turning effect about a fixed point or *Fulcrum*, and is measured by the product of the magnitude of the force and the *perpendicular distance* from the fulcrum to the line of action of the force (Figure 4.1).

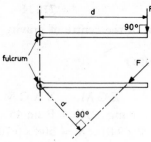

Fig. 4.1

$$\text{Moment} = \text{Force (N)} \times \text{distance (m)}$$
$$M = F \times d \text{ (Nm)}$$

Depending on the direction of the force, rotation could be *clockwise* or *anticlockwise* (Figure 4.2a).

Moments in Equilibrium If by the action of a number of forces a body is in equilibrium the moments must balance, otherwise the unbalanced resultant moment would cause the body to rotate (Figure 4.2b).

Principle of Moments When a number of coplanar forces act on a body and produce equilibrium, the sum of the clockwise moments taken about any point in the plane is equal to the sum of the anticlockwise moments about the same point.

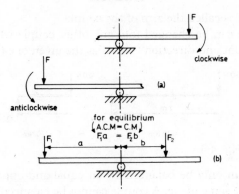

Fig. 4.2

For equilibrium

Σ Clockwise moments = Σ Anticlockwise moments

Σ C.M. = Σ A.C.M.

Or

The algebraic sum of the moments about any point in the plane is equal to zero.

Σ M = 0

(Σ is the Greek letter 'sigma' and means 'the sum of'.)

Couple A couple is formed by two equal unlike parallel forces which together produce or tend to produce rotation (Figure 4.3).

The moment of a couple is the product of one of the forces and the perpendicular distance between the lines of action of both forces.

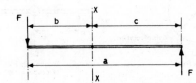

Fig. 4.3

Consider the light rod (Figure 4.3).

The resultant moment about axis XX $= (F \times b) + (F \times c)$
$$= Fb + Fc$$
$$= F(b + c)$$
$$= Fa$$

The moment of a couple $= F \times a (Nm)$

Distance 'a' is called the arm of the couple.

A couple can be changed into any other couple which has the same moment and direction (sense) as the given one (Figure 4.4).

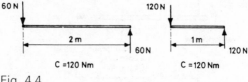

Fig. 4.4

A couple can only be balanced by an equal and opposite couple acting in the same plane. A couple cannot be balanced or replaced by a single force (Figure 4.5).

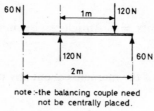

note :- the balancing couple need not be centrally placed.

Fig. 4.5

Torque   Like a couple, torque produces or tends to produce rotation. In the case of pulleys, shafts etc, this is obtained by a tangential force (F) which causes rotation about an axis (Figure 4.6).

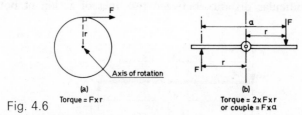

Fig. 4.6

The measurement of torque is the product of the applied tangential force and the perpendicular distance from the axis of rotation.

$$Torque = F \times r (Nm) \quad (r \text{ is the radius})$$

The units of couple and torque are the same as those used for moments (Nm).

Conditions of Equilibrium of Force Systems

1 *Two forces in equilibrium*
   a) Equal in magnitude but opposite in direction
   b) Colinear.
2 *Three forces in equilibrium*
   a) Coplanar.
   b) Lines of action of the forces parallel or concurrent.
   c) Algebraic sum of the moments of the forces about any point in the plane is zero ($\Sigma M = 0$).
3 *Four or more forces in equilibrium*
   a) Coplanar.
   b) Algebraic sum of the horizontal components of the forces is zero ($\Sigma F_H = 0$).
   c) Algebraic sum of the vertical components of the forces is zero ($\Sigma F_V = 0$).
   d) Algebraic sum of the moments of the forces about any point in the plane is zero ($\Sigma M = 0$).

NOTE: (*b*) and (*c*) of 3 could apply also to 2.

Worked Examples

1 a) If a force of 15 N is applied perpendicular to the edge of a door 0·8 m wide calculate the moment and state its sense.
   b) Calculate the magnitude of the force F_1 required at the mid-point of the door to prevent it turning.
   c) What would be the effect of force F_2 acting directly on the hinge?

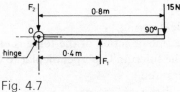

Fig. 4.7

a) *Clockwise moment* (C.M.) $= 15 \times 0·8$ Nm
$$= 12 \; Nm$$

b) As force F_1 acts at the mid-point of the door its distance from the hinge is 0·4 m.
   Anticlockwise moment (A.C.M.) $= F_1 \times 0·4$ Nm

For equilibrium

Take moments about the hinge

$$\Sigma \text{ A.C.M.} = \Sigma \text{ C.M.}$$
$$F_1 \times 0{\cdot}4 = 12$$
$$F_1 = 30 \text{ N}$$

The required force is 30 N

c) As F_2 acts directly on the hinge there is no turning effect (moment).

$$\text{Moment} = F_2 \times d$$
$$= F_2 \times 0 = 0$$

$\therefore$ *No turning effect.*

2   To illustrate two methods of finding the moment of a force when the line of action of the force is not perpendicular to the door.

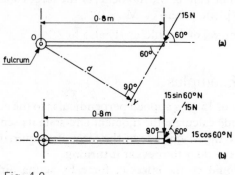

Fig. 4.8

Method 1 (by scale drawing)

(Figure 4.8a) The line of action of the force is extended and line OY drawn from the fulcrum O perpendicular to the line of action of the force is measured.

$$\text{Moment } (clockwise) = F \times d \text{ Nm}$$
$$= 15 \times 0{\cdot}69 \text{ Nm}$$
$$= 10{\cdot}4 \text{ Nm}$$

Method 2 (by rectangular components)

(Figure 4.8b) The vertical component of the 15 N force is 15 sin 60° and acts perpendicular to the door.

$$Moment\ (clockwise) = 15 \sin 60° \times 0.8\ Nm$$
$$= 13 \times 0.8\ Nm$$
$$= 10.4\ Nm$$

NOTE: The line of action of the horizontal component (15 cos 60°) acts along the door and through the hinge; therefore there is no moment.

3   A beam 5 m long rests horizontally on a wedge support as shown. Neglecting the mass of the beam calculate:
   a) The magnitude of the vertical force F required to maintain equilibrium.
   b) The magnitude and direction of the reaction at the wedge support.

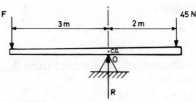

Fig. 4.9

Let the wedge be fulcrum O
Take moments about O
Anticlockwise moment (A.C.M.) = Clockwise moment (C.M.)

$$F \times 3 = 45 \times 2$$
$$F = 30\ N$$

Let the reaction at the wedge be R.
By action and reaction (Σ Upward forces = Σ downward forces):

$$R = 45\ N + 30\ N$$
$$= 75\ N$$

a) The required force is 30 N.
b) The reaction at the wedge support is 75 N vertically upwards.

4   AB is a uniform beam of length 4 m and mass 25 kg.
A mass of 30 kg is placed 0.6 m from A and another mass of 50 kg 1.4 m from B. Calculate the distance from A that a single support R should be placed to maintain equilibrium.

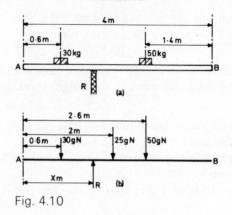

Fig. 4.10

(Figure 4.10b) Let R be the pillar support (reaction).
By action and reaction

$$R = (30g + 25g + 50g) \, N$$
$$R = 105g \, N$$

Let 'x' m be the distance of the support from A.
Take moments about A

$$\Sigma \, \text{A.C.M.} = \Sigma \, \text{C.M.}$$
$$R \times \text{'x'} = [(30g \times 0.6) + (25g \times 2) + (50g \times 2.6)]$$
$$105g \times x = 18g + 50g + 130g$$
$$x = \frac{198g}{105g}$$
$$x = 1.885$$

The support should be positioned 1·885 m from A.

5   A uniform beam AB, length 6 m and mass 50 kg rests horizontally
    on two supports A and B as shown in Figure 4.11.
    A force of 500 N acts perpendicular to the beam 2 m from A. Calcu-
    late the reactions at A and B.

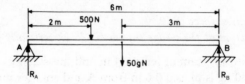

Fig. 4.11

As the beam is uniform the gravitational force will act at its mid-point.

By action and reaction

$$R_A + R_B = 500\,N + 50g\,N$$
$$R_A + R_B = 500\,N + 490\,N$$
$$R_A + R_B = 990\,N \quad . \quad . \quad . \quad . \quad . \quad .\textcircled{1}$$

Take moments about A
For equilibrium

$$\Sigma\text{ A.C.M.} = \Sigma\text{ C.M.}$$
$$R_B \times 6 = (500 \times 2) + (490 \times 3)$$
$$R_B = 412\,N$$

Take moments about B
For equilibrium

$$\Sigma\text{ C.M.} = \Sigma\text{ A.C.M.}$$
$$R_A \times 6 = (500 \times 4) + (490 \times 3)$$
$$R_A = 578\,N$$

The reaction at support A = 578 N
The reaction at support B = 412 N

NOTE: $R_A + R_B = 900\,N$. Check by addition $578 + 412 = 990$. R_B may be substituted in ① and R_A found, instead of taking moments about B.

6 A uniform rod (mass 10 kg), 4 m long is supported horizontally by two vertical ropes T_1 and T_2. If a mass of 5 kg is placed 0·2 m from the left hand end of the rod and a mass of 6 kg placed 0·4 m from T_2, calculate the magnitude of the force in each rope.

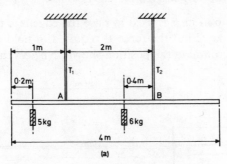

Fig. 4.12 (continued on next page)

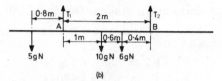

Fig. 4.12 (continued)

Take moments about A

$$\Sigma\,\text{A.C.M.} = \Sigma\,\text{C.M.}$$
$$(5g \times 0{\cdot}8) + (2 \times T_2) = (10g \times 1) + (6g \times 1{\cdot}6)$$
$$2T_2 = (10g + 9{\cdot}6g - 4g)\,\text{N}$$
$$T_2 = \frac{15{\cdot}6g}{2}\,\text{N}$$
$$T_2 = 7{\cdot}8g\,\text{N}$$

Take moments about B

$$\Sigma\,\text{C.M.} = \Sigma\,\text{A.C.M.}$$
$$T_1 \times 2 = (10g \times 1) + (6g \times 0{\cdot}4) + (5g \times 2{\cdot}8)$$
$$2T_1 = (10g + 2{\cdot}4g + 14g)\,\text{N}$$
$$T_1 = \frac{26{\cdot}4g}{2}\,\text{N}$$
$$T_1 = 13{\cdot}2g\,\text{N}$$

Check $\quad T_1 + T_2 = 7{\cdot}8g\,\text{N} + 13{\cdot}2g\,\text{N}$
$$= 21g\,\text{N}$$
$$= \text{the sum of the downward forces.}$$

The force in rope $T_1 = 129{\cdot}5$ N
The force in rope $T_2 = 76{\cdot}5$ N

7 A crowbar 1·6 m long is used to raise the edge of a uniform crate
   of mass 100 kg. Find the force F required to hold the crate just
   clear of the ground in the position shown. Neglect the mass of the
   crowbar.

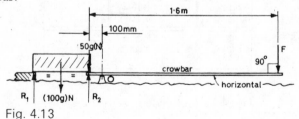

Fig. 4.13

As the crate is uniform $R_1 = R_2 = 50g = 490\,N$
Take moments about O
For equilibrium

$$\Sigma\ C.M. = \Sigma\ A.C.M.$$

$$F \times 1{\cdot}5 = 490 \times \frac{100}{1\,000} \quad \begin{bmatrix} Note : \text{ It is important that} \\ \text{units are consistent} \end{bmatrix}$$

$$F = \frac{490 \times 0{\cdot}1}{1{\cdot}5}\,N$$

$$F = 32{\cdot}6\,N$$

The required force is 32·6 N

NOTE: A *lever* is a rigid bar which is free to rotate about a point called the *fulcrum*. Levers are found in many forms; some examples are: crowbar, wheelbarrow, lever safety valve, nut crackers, guillotine, oar of a rowing boat, etc.

8   A guillotine is used to cut sheets of thin cardboard. If the arm of the guillotine is 350 mm long and the force applied at the handle is 30 N calculate the resistance of the cardboard for the position shown.

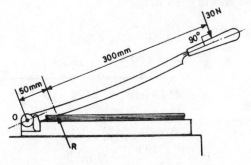

Fig. 4.14

Let R be the resistance of the cardboard.
Take moments about the fulcrum O.
For equilibrium

$$\Sigma\ A.C.M. = \Sigma\ C.M.$$

$$R \times 50 = 30 \times 350$$

$$R = 210\,N$$

The resistance of the cardboard is 210 N.

9   Figure 4.15 shows a drawing of a lever safety valve mechanism. The mass of the counterbalance is 8 kg and the diameter of the valve is 60 mm. Calculate the maximum pressure within the boiler to operate the valve. Neglect the mass of the valve and lever.

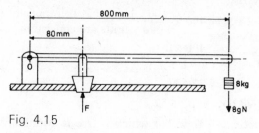

Fig. 4.15

To find the force F acting on the valve
Take moments about O
For equilibrium

$$\Sigma \text{ A.C.M.} = \Sigma \text{ C.M.}$$
$$F \times 80 = 8 \times 9.81 \times 800$$
$$F = 785 \text{ N}$$

To find the area of the valve

$$\text{Area} = \frac{\pi d^2}{4}$$

$$A = \frac{3.14 \times 60 \times 60}{4} \text{ mm}^2$$

$$A = 2\,826 \text{ mm}^2 = 2\,826 \times 10^{-6} \text{ m}^2$$

$$\text{Pressure} = \frac{\text{Force}}{\text{Area}}$$

$$P = \frac{785}{2\,826 \times 10^{-6}} \text{ N/m}^2$$

$$P = 0.278 \times 10^6 \text{ N/m}^2$$

$$P = 278 \text{ kN/m}^2$$

Maximum pressure is 278 kN/m².

10   In the compound lever system shown a mass of 25 kg is placed at C and is supported by a force F applied vertically at B. The lever AB pivots about O_2 and lever CD pivots about O_1. Calculate:
   a) The magnitude and the nature of the force in the link GH.
   b) The magnitude of the reactions at O_1 and O_2.

c) The magnitude of the force F.

NOTE: A compound lever system is made up of a number of simple levers linked together by rods (links). The joints are assumed pin-jointed. The system allows a reduction in space required and is of greater advantage to the operator.

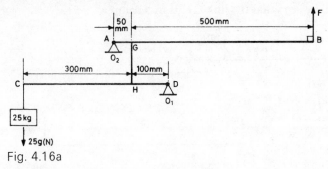

Fig. 4.16a

Treat CD as a simple lever (Figure 4.16b).

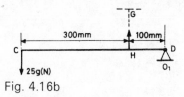

Fig. 4.16b

The direction of the force in link GH at H must act upwards.
Take moments about O_1
For equilibrium

$$\Sigma \text{ C.M.} = \Sigma \text{ A.C.M.}$$

$$\text{Force in link GH} \times 100 = (25g \times 400)$$
$$\text{GH} = 100 \times 9 \cdot 81 \text{ N}$$
$$\text{GH} = 981 \text{ N}$$

Consider the second simple lever AB (Figure 4.16c).

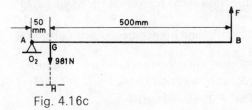

Fig. 4.16c

Take moments about O_2
For equilibrium

$$\Sigma \text{ A.C.M.} = \Sigma \text{ C.M.}$$
$$F \times 550 = 981 \times 50$$
$$F = 89 \text{ N}$$

It will be seen that the link GH is in tension.

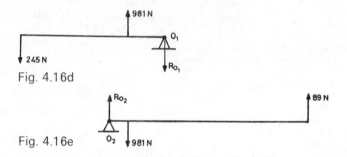

Fig. 4.16d

Fig. 4.16e

To find the reaction at O_1 and O_2
By action and reaction

(Figure 4.16d) Reaction O_1 = 981 N − 245 N
RO_1 = 736 N ↓
(Figure 4.16e) Reaction O_2 = 981 N − 89 N
RO_2 = 892 N ↑

a) *The magnitude of the force in the link GH is 981 N and the link is in tension.*
b) *Magnitude of reaction O_1 is 736 N ↓*
   Magnitude of reaction O_2 is 892 N ↑
   The magnitude of the force F is 89 N

11 The bell crank lever shown is used as part of a railway signal system. Calculate:
    a) The force F required to maintain equilibrium.
    b) The magnitude and direction of the reaction at the hinge pin O.

Resolving the 25 N force parallel and perpendicular to the arm AO.
Take moments about O
For equilibrium

$$\Sigma \text{ C.M.} = \Sigma \text{ A.C.M.}$$
$$F \times 0.3 \text{ m} = 25 \sin 60° \times 0.4 \text{ m}$$

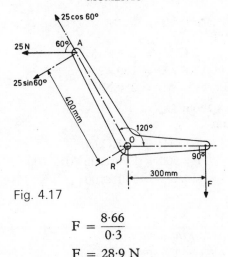

Fig. 4.17

$$F = \frac{8 \cdot 66}{0 \cdot 3}$$

$$F = 28 \cdot 9 \text{ N}$$

The reaction at the hinge can be found by one of the methods used in chapters 2 or 3 (parallelogram or triangle of forces).

a) *The force F required is 28·9 N*

b) *The magnitude of the reaction at the hinge pin O is 38·3 N acting* ∠49°

NOTE: Bell crank levers are used to change the direction of forces.

12 A simple brake mechanism consists of a simple lever and a right angled bell crank lever connected by a link. Calculate the vertical force F required at the handle to exert a force of 200 N on the wheel at P. Neglect the mass of the levers and link.

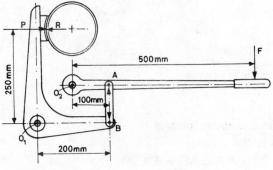

Fig. 4.18

Take moments about O_1
For equilibrium

$$\Sigma \text{ C.M.} = \overset{\leftarrow}{\Sigma \text{ A.C.M.}}$$
$$\downarrow \text{Force in AB} \times 200 = 200 \times 250$$
$$\text{Force in AB} = 250 \text{ N}$$

Take moments about O_2
For equilibrium

$$\Sigma \text{ C.M.} = \Sigma \text{ A.C.M.}$$
$$F \times 500 = \uparrow \text{Force in AB} \times 100$$
$$F = \frac{250 \times 100}{500} \text{ N}$$
$$F = 50 \text{ N}$$

The force required at the handle is 50 N.
The link AB is in compression.

13 A system of levers is part of a machine mechanism. The fulcrums are at B and D. The members are pin-jointed at C, E and F. A force of 77 N is applied at A as shown. Neglect the mass of the levers and links. Determine:

a) The magnitude and nature of the force in member FG.
b) The magnitude and direction of the reaction at fulcrum D.

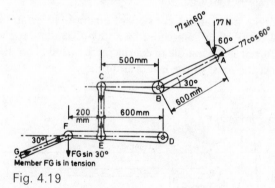

Fig. 4.19

Take moments about fulcrum B
For equilibrium

$$\Sigma \text{ A.C.M.} = \Sigma \text{ C.M.}$$
$$\downarrow \text{ Force in CE} \times 500 = 77 \sin 60° \times 600$$

$$\text{Force in CE} = \frac{77 \times 0.866 \times 600}{500} \text{ N}$$

$$= 80 \text{ N}$$

Take moments about fulcrum D
For equilibrium

$$\Sigma \text{ A.C.M.} = \Sigma \text{ C.M.}$$
$$\text{FG sin } 30° \times 800 = \uparrow \text{Force in CE} \times 600$$
$$400 \text{ FG} = 80 \times 600 \text{ N}$$
$$\text{FG} = \frac{80 \times 600}{400} \text{ N}$$
$$= 120 \text{ N}$$

a) *The force in the member FG is 120 N.*
The member is in tension.

The reaction at fulcrum D can be found either graphically or analytically using the triangle of forces.

b) *Reaction at fulcrum D is 104 N acting* $11°$ ◢◣

14 A light rod AB hinged at B is maintained in a horizontal position by means of a spring at C. If a mass of 12 kg is placed at end A of the rod calculate:
a) The force acting on the spring.
b) The reaction at hinge pin B.
c) The extension of the spring if the spring stiffness is 80 N/mm.
Neglect the mass of the rod.

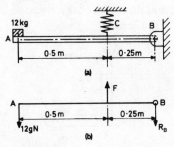

Fig. 4.20

(Figure 4.20b) Let F be the force in the spring
Take moments about B

For equilibrium

$$\Sigma \text{ C.M. } = \Sigma \text{ A.C.M.}$$
$$F \times 0\cdot 25 = 12\,g \times 0\cdot 75$$
$$F = \frac{12 \times 9\cdot 81 \times 0\cdot 75}{0\cdot 25}\,N$$
$$F = 353\,N$$

By action and reaction

$$12\,g + R_B = 353\,N$$
$$R_B = 353\,N - (12 \times 9\cdot 81)\,N$$
$$R_B = 235\,N\downarrow$$

$$\text{Displacement of spring (mm)} = \frac{\text{Force (N)}}{\text{Spring stiffness (N/mm)}}$$
$$= \frac{353}{80}\,mm$$
$$= 4\cdot 4\,mm$$

a) The force in the spring is 353 N.
b) The reaction at hinge pin B is 235 N↓.
c) The extension of the spring is 4·4 mm.

15 A right angled bell crank lever AOB, pivoted at O rests on a spring so that OB is horizontal. The spring stiffness is 10 N/mm. If a force of 20 N is now applied at A as shown calculate:
a) The force acting on the spring.
b) The distance that the spring will be compressed.
c) The magnitude and direction of the reaction at the pivot pin O.
Neglect the mass of the lever.

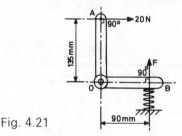

Fig. 4.21

Let F be the force in the spring
Take moments about O

For equilibrium

$$\Sigma \text{ A.C.M.} = \Sigma \text{ C.M.}$$
$$F \times 90 = 20 \times 135$$
$$F = \frac{2700}{90} \text{ N}$$
$$F = 30 \text{ N}$$

$$\text{Displacement of spring} = \frac{\text{Force}}{\text{Spring stiffness}}$$
$$= \frac{30}{10} \text{mm}$$
$$= 3 \text{ mm}$$

Reaction at pivot pin O

$R_0 = \sqrt{30^2 + 20^2}$ N   Applied force = 20 N

$R_0 = \sqrt{1300}$ N   Reaction at spring = 30 N

$R_0 = 36$ N   Reaction at pivot pin O $= R_0$

a) *The force in the spring is 30 N.*
b) *The spring is compressed 3 mm.*
c) *The reaction at the pivot pin O is 36 N acting* 56·3°

16 A handwheel (380 mm diameter) on the deck of an oil tanker is used to operate valves controlling the flow of oil through pipes. A seaman exerts a force of 50 N with each hand to turn the handwheel. Calculate the torque on the lead shaft.

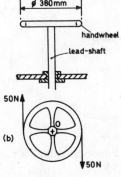

Fig. 4.22

Consider Figure 4.22b.

Take moments about O.

This consists of two clockwise moments.

Turning effect of the wheel

$$= [(50 \times 190) + (50 \times 190)] \text{ Nmm}$$
$$= 9\,500 \text{ Nmm} + 9\,500 \text{ Nmm}$$
$$= 19\,000 \text{ Nmm}$$
$$= 19 \text{ Nm (clockwise)}$$

This turning effect is commonly referred to as the torque on the shaft.

or Torque $= 2 \times F \times r$ F = the applied force.
$\quad\quad\quad\quad = 2 \times 50 \times 190 \text{ Nmm}$ r = the radius of the hand-
$\quad\quad\quad\quad = 100 \times 190 \text{ Nmm}$ wheel.
$\quad\quad\quad\quad = 19\,000 \text{ Nmm}$
$\quad\quad\quad\quad = 19 \text{ Nm}$

or Couple $= F \times a$ F = the applied force.
$\quad\quad\quad\quad = 50 \times 380 \text{ Nmm}$ a = the arm of the couple
$\quad\quad\quad\quad = 19\,000 \text{ Nmm}$ (i.e. diameter of the
$\quad\quad\quad\quad = 19 \text{ Nm}$ handwheel).

The torque on the lead shaft = 19 Nm

17   A pulley 300 mm diameter is keyed to a shaft 40 mm diameter. If a force of 300 N is applied tangential to the pulley, calculate:

a) The torque on the pulley shaft.

b) The force F acting on the key.

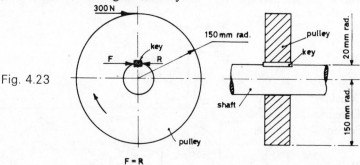

Fig. 4.23

a) Torque on shaft = Force on pulley × radius of pulley
$\quad\quad\quad\quad\quad\quad = 300 \text{ N} \times 150 \text{ mm}$
$\quad\quad\quad\quad\quad\quad = 45\,000 \text{ Nmm}$
$\quad\quad\quad\quad\quad\quad = 45 \text{ Nm}$

b) Torque on shaft = Force on key × radius of shaft

$$45\,\text{Nm} = F \times 20 \times 10^{-3}\,\text{Nm}$$

$$F = \frac{45}{0\cdot02}\,\text{N}$$

$$F = 2\,250\,\text{N} = 2\cdot25\,\text{kN}$$

a) *Torque on pulley shaft is 45 Nm.*
b) *Force on key is 2·25 kN.*

18 A truck with crane attachment is shown in Figure 4.24. Details are as follows:
Truck: mass 3×10^3 kg. Centre of gravity 1 m to rear of front axle.
Crane: mass 2×10^3 kg. Centre of gravity directly above rear axle.
Load carried by crane 500 kg.
The centre of gravity of truck, crane and load act on the longitudinal centre line. Calculate:

a) The force on each axle.
b) The reaction between each wheel and the ground (two wheels on front axle and four wheels on rear axle).

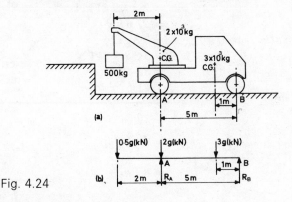

Fig. 4.24

Consider Figure 4.24b.
Take moments about A
For equilibrium

$$\Sigma\,\text{A.C.M.} = \Sigma\,\text{C.M.}$$
$$(R_B \times 5) + (0\cdot5\,g \times 2) = 3\,g \times 4$$
$$5R_B = 12\,g - 1\,g$$

$$R_B = \frac{11\,g}{5}\,\text{kN}$$

$$R_B = 2\cdot2\,g \ \text{kN}$$

By action and reaction

$$R_A + R_B = (0\cdot5\,g + 2\,g + 3\,g)\,\text{kN}$$
$$R_A + R_B = 5\cdot5\,g\,\text{kN}$$
$$R_A = (5\cdot5\,g - 2\cdot2\,g)\,\text{kN}$$
$$R_A = 3\cdot3\,g \ \text{kN}$$

Check: Take moments about B
For equilibrium

$$\Sigma\ \text{C.M.} = \Sigma\ \text{A.C.M.}$$
$$R_A \times 5 = [(3\,g \times 1) + (2\,g \times 5) + (0\cdot5\,g \times 7)]$$
$$5R_A = (3\,g + 10\,g + 3\cdot5\,g)$$
$$R_A = \frac{16\cdot5\,g}{5}\text{kN}$$
$$R_A = 3\cdot3\,g\,\text{kN}$$

a) *Force on the front axle* $= 2\cdot2\,g\,\text{kN}$
                                    $= 2\cdot2 \times 9\cdot81\,\text{kN}$
                                    $= 21\cdot6\,kN$

    Force on the rear axle $= 3\cdot3\,g\,\text{kN}$
                                      $= 3\cdot3 \times 9\cdot81\,\text{kN}$
                                      $= 32\cdot4\,kN$

b) *Reaction at each front wheel* $= \dfrac{21\cdot6}{2}\text{kN}$
                                          $= 10\cdot8\,kN$

    Reaction at each rear wheel $= \dfrac{32\cdot4}{4}\ \text{kN}$
                                          $= 8\cdot1\,kN$

19   The outline of an articulated vehicle is shown in Figure 4.25. The
motor unit (mass 2 tonne) has its centre of gravity 0·6 m to the
rear of the front axle A. The trailer and its contents (mass 3
tonne) has its centre of gravity 1 m to the front of the rear axle C.
The centre of gravity of the vehicle and loaded trailer act through
the longitudinal centre line.
Determine:
a) The force at the coupling pin P.
b) The force on each of the three axles A, B and C.

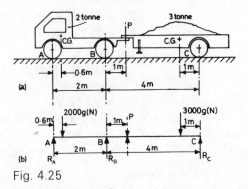

Fig. 4.25

Consider Figure 4.25b
Take moments about axle C
For equilibrium

$$\Sigma \text{ C.M.} = \Sigma \text{ A.C.M.}$$
$$P \times 3 = 3\,000\,g \times 1$$
$$P = 1\,000\,g\,N$$

Take moments about axle A
For equilibrium

$$\Sigma \text{ A.C.M.} = \Sigma \text{ C.M.}$$
$$R_B \times 2 = (2\,000\,g \times 0.6) + (P \times 3)$$
$$2R_B = 1\,200\,g + (1\,000\,g \times 3)$$
$$R_B = \frac{1\,200\,g + 3\,000\,g}{2}$$
$$R_B = 2\,100\,g\,N$$

Take moments about axle B
For equilibrium

$$\Sigma \text{ C.M.} = \Sigma \text{ A.C.M.}$$
$$(R_A \times 2) + (P \times 1) = (2\,000\,g \times 1.4)$$
$$2R_A + 1\,000\,g = 2\,800\,g$$
$$R_A = \frac{2\,800\,g - 1\,000\,g}{2}$$
$$R_A = 900\,g\,N$$

By action and reaction

$$R_A + R_B + R_C = mg\,N$$
$$900\,g\,N + 2\,100\,g\,N + R_C = 5\,000\,g\,N$$

$$R_C = (5\,000\,g - 3\,000\,g)\,N$$
$$R_C = 2\,000\,g\,N$$

a) *Force exerted at coupling P is 1 000 g N*
$$= 1\,000 \times 9{\cdot}81\,N$$
$$= 9\,810\,N$$
$$= 9{\cdot}81\,kN$$

b) *Force on axle A $= 900\,g\,N = 900 \times 9{\cdot}81 = 8\,830\,N = 8{\cdot}83\,kN$*
Force on axle B $= 2\,100\,g\,N = 2\,100 \times 9{\cdot}81 = 20\,600\,N = 20{\cdot}6\,kN$
Force on axle C $= 2\,000\,g\,N = 2\,000 \times 9{\cdot}81 = 19\,620\,N = 19{\cdot}62\,kN$

20   The outline of a crane is shown.
Calculate for the position shown in Figure 4.26:
a) The vertical reactions at each set of wheels A and B.
b) The mass that will just cause the crane to become unstable
and the reactions for this condition.

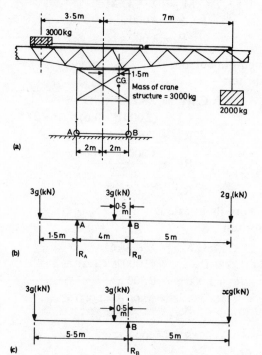

Fig. 4.26

a) (Figure 4.26b) By action and reaction

$$R_A + B_B = (3\,g + 3\,g + 2\,g)\,kN$$
$$R_A + R_B = 8\,g\,kN \quad . \quad . \quad . \quad . \quad . \quad . \quad . \quad \text{①}$$

Take moments about A
For equilibrium

$$\Sigma \text{ A.C.M.} = \Sigma \text{ C.M.}$$
$$(R_B \times 4) + (3\,g \times 1\cdot5) = (2\,g \times 9) + (3\,g \times 3\cdot5)$$
$$4R_B = (18\,g + 10\cdot5\,g - 4\cdot5\,g)\,kN$$
$$R_B = \frac{24\,g}{4}\,kN$$
$$R_B = 6\,g\ kN$$
$$\text{From ①} \quad R_A + R_B = 8\,g\ kN$$
$$R_A + 6\,g\,kN = 8\,g\ kN$$
$$R_A = 2\,g\ kN$$

The reaction at A = 2g kN = 19·6 kN
The reaction at B = 6g kN = 59 kN

b) (Figure 4.26c) By action and reaction

$$R_B = (3\,g + 3\,g + x\,g)\,kN$$
$$R_B = (6\,g + x\,g)\,kN \quad . \quad . \quad . \quad . \quad . \quad . \quad \text{②}$$

Note that the reaction at A will be zero for this condition.
i.e. $R_A = 0$

Take moments about B
For equilibrium

$$\Sigma \text{ C.M.} = \Sigma \text{ A.C.M.}$$
$$(x\,g \times 5) = (3\,g \times 5\cdot5) + (3\,g \times 0\cdot5)$$
$$5\,x = 16\cdot5\,g + 1\cdot5\,g$$
$$x = \frac{18}{5}$$
$$x = 3\cdot6$$

The limiting mass for stability = 3·6 × 10³ kg = 3 600 kg

Reactions $\quad\quad R_A = 0$
From ② $\quad\quad\quad R_B = (6\,g + x\,g)\,kN$
$$R_B = (6 + 3\cdot6\,g)\,kN$$
$$R_B = 9\cdot6\,g\ kN$$
$$R_B = 94\cdot3\,kN$$

The reaction at A = 0
The reaction at B = 94·3 kN

Examples

1   Figure 4.27:

    a) A metal bar 700 mm long rests in a horizontal position on a wedge as shown. Neglect the mass of the bar. Calculate:

        i) the magnitude of the force F to maintain equilibrium.

        ii) the magnitude and direction of the reaction at the wedge.

    b) A length of wood suspended from a rope is kept in a horizontal position due to the action of the loads shown. Neglect the mass of the wood. Calculate:

        i) distance X to maintain equilibrium.

        ii) the tension in the rope.

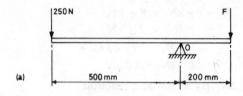

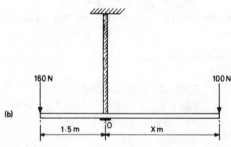

Fig. 4.27

2   Figure 4.28 represents a metre stick, horizontal and with loads of 12 N and 8 N placed as shown. Where must a single support be placed for balance? (Calculate distance X.) Neglect the mass of the metre stick.

Fig. 4.28

3 A uniform plank 5 m long is placed on top of a low wall in such a way that one end projects 0·9 m at right angles from the wall. If a mass of 48 kg is placed at this end so that the plank is horizontal, what is the mass of the plank?

4 A uniform bar 800 mm long and of mass 8 kg is suspended by means of a cord at its mid-point. Three tubular pieces of metal each of mass 2 kg are slipped onto the bar and rest in the following positions:— one 200 mm from the left hand end; one 150 mm from the right hand end. If the bar is to balance horizontally calculate:
a) The position of the third tubular piece of metal on the bar.
b) The tension in the cord.

5 A beam AB is horizontal and rests on supports at A and B. A mass of 10 kg is placed 0·2 m from A and a mass of 7 kg is placed 0·5 m from B. Neglect the mass of the beam. Calculate the magnitude of the support reactions at A and B (Figure 4.29).

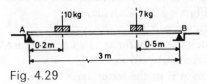

Fig. 4.29

6 AC represents a uniform beam of mass 204 kg supported at A and B in a horizontal position. If two forces, each of 3 kN act as shown (Figure 4.30) calculate the reactions at the supports.

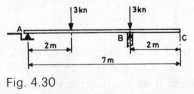

Fig. 4.30

7 A uniform foot bridge AB of mass 408 kg and length 5 m is used as a temporary means across a stream. Wooden supports at each end A and B maintain the bridge in a horizontal position. Calculate the reactions at each support if a man of mass 78 kg stands on the bridge 2 m from end A.

8 In an engineering workshop a uniform machine shaft AB of length 2·5 m rests with end A on the edge of a metal platform

and is held in a horizontal position by a vertical rope sling attached 800 mm from end B. If the tension in the rope sling is 11 kN calculate:

a) The mass of the machine shaft.

b) The reaction at the platform.

9 A wheel and lever are keyed to a shaft which is carried in smooth bearings. Calculate the force F which will balance a force of 100 N acting at the end of the lever as shown (Figure 4.31). Neglect the mass of the lever.

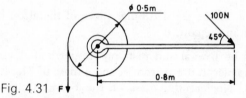

Fig. 4.31 F

10 A bar AB of mass 10 kg is pivoted at B. A force of 30 N is applied as indicated (Figure 4.32). Calculate:

a) The moments about B.

b) By how much the 30 N force must be increased to balance the bar in a horizontal position.

c) The reaction at B in the latter case.

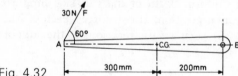

Fig. 4.32

11 A guillotine is used to cut strips of cardboard. If the arm of the guillotine is 300 mm long and the cardboard offers a resistance of 100 N when 75 mm from the fulcrum calculate the force required when applied perpendicular to the handle.

12 To cut a small piece of tinplate using snips, a force of 50 N is required. If the tinplate is 20 mm from the fulcrum and the applied force 150 mm from the fulcrum, calculate the resistance offered by the tinplate.

13 The lever safety valve mechanism shown in Figure 4.33 has the following particulars:

Boiler pressure: $633 \, kN/m^2$. Mass of counterbalance 9 kg.
Dimensions as shown in Figure 4.33.
Calculate the diameter of valve A.

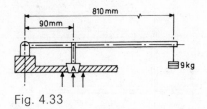

Fig. 4.33

14 Figure 4.34:
   a) A right angled bell crank lever is pivoted at O.
      For a pull of 40 N on arm A as shown calculate:
      i) the force F required to maintain equilibrium.
      ii) the magnitude and direction of the reaction at pivot pin O.
   b) The bell crank lever shown is used to change the direction of a
      200 N force. Calculate:
      i) the distance X to maintain equilibrium.
      ii) the magnitude and direction of the reaction at pivot pin O.

   NOTE: The reactions at the pivot pins may be found by graphical
   means.

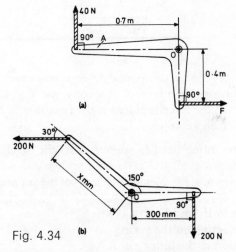

Fig. 4.34

15 A simple braking system used in a light industrial machine is
   shown (Figure 4.35). A force of 40 N applied at A is sufficient to

stop the disc B revolving. Calculate the force F required to stop the disc and the magnitude and direction of the reactions at O_1 and O_2. Neglect the mass of levers and link.

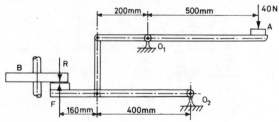

Fig. 4.35

16 A diagrammatic drawing of a weighing machine is shown. If a force of 90 N is applied at A, find the mass 'm' which can be centrally supported on platform 'XX' (Figure 4.36) to maintain equilibrium. What is the magnitude of the force acting on pivot O_2?

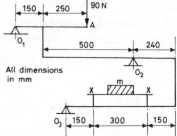

Fig. 4.36

17 On an assembly line a compound lever arrangement is used to knock components from level X to a conveyor Y at a lower level (Figure 4.37). If the force required at the handle is 70 N, calculate the force acting on a component.

18 A resilient mounting used to support a fragile machine of mass 16·3 kg, is hinged at A and rests on a spring at S so that AB is horizontal (Figure 4.38). Neglect the mass of the support structure. Calculate:
a) The force in F_1 and F_2.
b) The force acting on the spring.
c) The reaction at the hinge pin A.
d) The distance point B will rise if the mass is removed, if the spring stiffness is 11 N/mm.

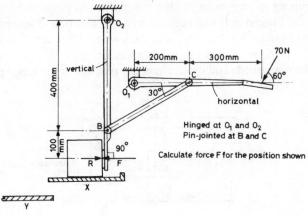

Fig. 4.37

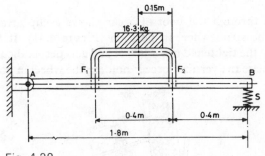

Fig. 4.38

19 A right angled bell crank lever pivoted at O, rests on a cam at B and against a spring A which is compressed 2 mm for the position shown in Figure 4.39. If the cam turns through an angle of 90° and the spring stiffness is 5 N/mm calculate:

a) The total distance the spring is compressed.
b) The total force acting on the spring.
c) The force at B acting on the cam.

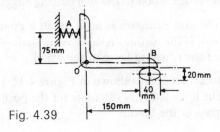

Fig. 4.39

20   A clamping arrangement holds a component in position by means
     of a lever hinged at B and secured by a nut and bolt (Figure 4.40).
     The clamping nut exerts a force of 2 kN when tightened. Calculate:
     a) The force F exerted on the component.
     b) The magnitude and direction of the reaction at the hinge pin B.

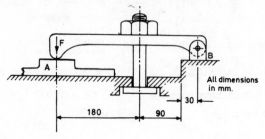

All dimensions
in mm.

Fig. 4.40

21   A section through the boot of a car shows a clip arrangement
     holding the spare wheel in position (Figure 4.41). If the force
     exerted by the tightened nut is 800 N and acts perpendicular to the
     clip, calculate the vertical force F holding the wheel in position.

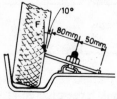

Fig. 4.41

22   In a power station an engineer exerts a force of 60 N with each
     hand on the handwheel of a valve control shaft. Calculate the
     torque on the control shaft if the handwheel is 400 mm in diameter.

23   A wood drill 20 mm dia is subjected to a torque of 5·6 Nm. Cal-
     culate the reaction of the wood to the two cutting edges.

24   A rotary cutter 160 mm diameter is used to cut a groove along a
     length of hardwood. If the torque on the driving shaft of the cutter
     is 48 Nm calculate the resisting force of the wood.

25   A lever is keyed to a shaft at A as shown in Figure 4.42. If a vertical
     force applied to the lever is 200 N calculate for the position shown
     (neglecting the mass of the lever):

a) The torque on the shaft.
b) The force on the key.

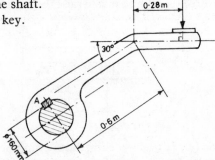

Fig. 4.42

26 A simple engine mechanism is shown in Figure 4.43. If the force on the piston is 5 kN determine for the position shown:
a) The force in the connecting rod BC.
b) The cylinder wall reaction at E.
c) The torque on the crank shaft at A.

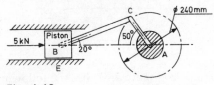

Fig. 4.43

27 A break-down vehicle has a hoisting arrangement consisting of an arm supported by a single hydraulic cylinder. For the position shown in Figure 4.44 calculate:
a) The force acting on the cylinder.
b) The pressure in the cylinder if the cylinder is 80 mm diameter.

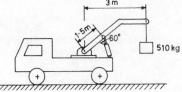

Fig. 4.44

28 A wagon of mass 2×10^3 kg and contents of mass 3×10^3 kg have together a common centre of gravity positioned as shown (Figure 4.45). Calculate:
a) The force acting on each of the wagon's four wheels (assume the C. of G. acts through the longitudinal axis of the wagon).

b) The force F required just to raise the wheels at B off the ground.

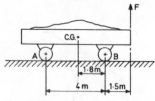

Fig. 4.45

29 A tractor (C) is used to carry farm equipment (D) as shown in Figure 4.46. If the mass of the tractor is 3 tonne and the mass of the equipment 500 kg calculate the force on each axle A and B.

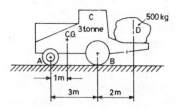

Fig. 4.46

30 An articulated vehicle (Figure 4.47) has the following particulars: Mass of cab 2 tonne; mass of trailer 1 tonne; mass of load 5 tonne; position of the C. of G. of the cab is 1 metre from connection X; position of the centre of gravity of the trailer and load is 2 metres in front of the trailer's wheels. Calculate:
a) The force at connection X.
b) The force on each axle at A, B and C.

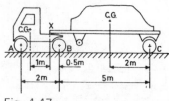

Fig. 4.47

31 A crate AB containing a machine part rests horizontally with end A on the end of the wagon and end B supported by a rope (F) as shown in Figure 4.48. The mass of the crate and the machine part is 1 020 kg, and the mass of the wagon is 2 040 kg. The position of each centre of gravity is shown. Calculate:

a) The force in rope F.
b) The force acting on the wagon at A.
c) The force acting on each axle of the wagon at C and D.

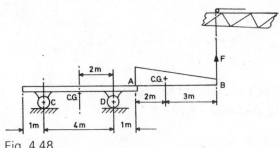

Fig. 4.48

32 The crate and machine part in question 31 is now lifted clear of
    the wagon and held by means of a rope from a crane in the position
    shown in Figure 4.49. For the particulars given (assuming that the
    centres of gravity are on the axis of symmetry):
    a) Calculate the reactions at A and B to each of the four wheels.
    b) Calculate the maximum load the crane can lift for a minimum
       reaction of 2 kN to one wheel at B.
    c) If the crane has 'no load' state the position of the counter-
       balance ⓒ so that the reactions at A and B are the same.

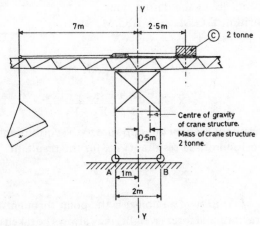

Fig. 4.49

5
Centre of Gravity

To find the position of the resultant of a number of parallel forces

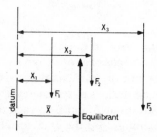

Fig. 5.1

The resultant of the forces is $(F_1 + F_2 + F_3)$. In order to find the position where it acts, the forces can be balanced by the equilibrant. Let the equilibrant act at a distance $\overline{X}$ from the datum.

Take moments about the datum line.

For equilibrium, Σ C.M. = Σ A.C.M.

i.e. $F_1x_1 + F_2x_2 + F_3x_3 = EQ \times \overline{X}$.

$$\therefore \quad \overline{X} = \frac{F_1x_1 + F_2x_2 + F_3x_3}{EQ}$$

$$\overline{X} = \frac{F_1x_1 + F_2x_2 + F_3x_3}{F_1 + F_2 + F_3}$$

Since the resultant is equal but opposite to the equilibrant this also gives the position of the resultant; *i.e.* for the resultant:

$$\overline{X} = \frac{\Sigma \, Fx}{\Sigma \, F}$$

The *Centre of Gravity* of an object is the point through which the resultant gravitational force (weight) may always be taken as acting, no matter in what position the object is placed, *i.e.* the point at which it balances. To specify the centre of gravity completely, three dimensions x, y and z (Figure 5.2) must be given. In some cases, owing to the symmetry of the object, it may be necessary to calculate only one or two. In most cases it is possible to work with volumes instead of weights, if the assumption is made that the object is of uniform material throughout.

i.e. $\text{W} \propto \text{V}$

If the object is also of uniform thickness, then:

$$\text{W} \propto \text{A}$$

and areas may be used instead of weights.

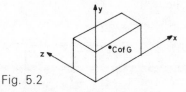

Fig. 5.2

The technique of calculating centres of gravity is to use the same method as for the resultant of parallel forces. This is done in each of the three directions in turn. In all except the simplest cases it is an advantage to tabulate the calculations.

Centroid   If we consider a very thin sheet or lamina, the centre of gravity of this may be considered as being on the surface. This point will also be the centre of area of the lamina. The centre of area is known as the *Centroid*. Centroids are of particular importance in structural mechanics. The centroid of a beam section is the point where there is no stress; *i.e.* to one side the beam is in tension and to the other side the beam is in compression.

Centroids are also required for finding the thrust on plates immersed in liquids, *e.g.* lock gates. Below (Figure 5.3) are shown some of the most commonly used geometrical shapes, along with methods for finding their centroids.

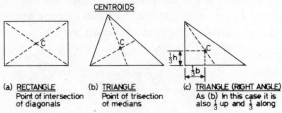

CENTROIDS

(a) RECTANGLE
Point of intersection
of diagonals

(b) TRIANGLE
Point of trisection
of medians

(c) TRIANGLE (RIGHT ANGLE)
As (b) In this case it is
also $\frac{1}{3}$ up and $\frac{1}{3}$ along

Fig. 5.3 (continued on next page)

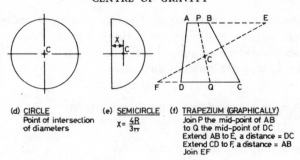

(d) <u>CIRCLE</u>
Point of intersection
of diameters

(e) <u>SEMICIRCLE</u>
$x = \frac{4R}{3\pi}$

(f) <u>TRAPEZIUM (GRAPHICALLY)</u>
Join P the mid-point of AB
to Q the mid-point of DC
Extend AB to E, a distance = DC
Extend CD to F, a distance = AB
Join EF

Fig. 5.3 (continued)

<u>CENTRES OF GRAVITY</u>

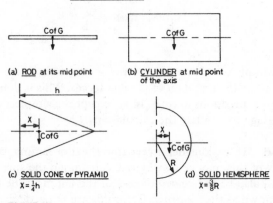

(a) <u>ROD</u> at its mid point

(b) <u>CYLINDER</u> at mid point
of the axis

(c) <u>SOLID CONE or PYRAMID</u>
$x = \frac{1}{4}h$

(d) <u>SOLID HEMISPHERE</u>
$x = \frac{3}{8}R$

Fig. 5.4

Stability   Figure 5.5 shows a cone in various positions. If the cone (a) is tilted slightly it will return to its original position. This is a state of *Stable Equilibrium*. Sketch (b) shows *Unstable Equilibrium*, where, if the cone is tilted slightly, it will continue to fall over. *Neutral Equilibrium* is shown in (c). In this case, if moved, the cone will roll slightly then come to rest. Notice that the centre of gravity remains at the same height throughout when in neutral equilibrium. For position (a) the cone will become unstable when the weight acts outside the base as shown at (d).

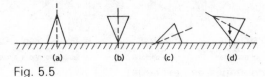

(a)   (b)   (c)   (d)

Fig. 5.5

To aid stability it is important to have:
a) as wide a base as possible
b) as low a centre of gravity as possible.

The question of stability is of importance when considering road vehicles: *e.g.* a racing car has a wide wheelbase and a low centre of gravity which enable it to withstand high centrifugal forces when cornering at high speeds. A double deck bus can be made more stable by designing it with the centre of gravity as low as possible (Figure 5.6).

Fig. 5.6

 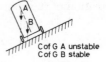

Worked Examples

1 (Figure 5.7) A uniform shaft of length 30 cm has a mass of 2 kg. Two pulleys, each of thickness 4 cm and of masses 3 kg and 1 kg respectively are fixed, one at each end as shown. The shaft has to be supported on one bearing placed at the centre of gravity of the assembled shaft and pulleys. Find the position of the centre of the bearing.

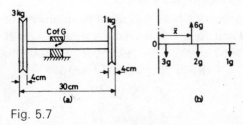

Fig. 5.7

Since the shaft is uniform the centre of gravity will lie on the shaft axis.

Suppose the shaft to be supported at its C. of G. The upward force required here would be equal to the sum of the gravitational forces on the shaft (Figure 5.7b).

Taking moments about O

$$(\text{Total mass} \times g) \times \overline{X} = (2g \times 15) + (3g \times 2) + (1g \times 28)$$

(*i.e.* Moment of whole = Sum of the moments of the parts)

$$\therefore \quad 6 \times \overline{X} = 30 + 6 + 28$$

$$\overline{X} = \frac{64}{6}$$

$$= 10\cdot67$$

∴ *Centre of the bearing must lie 10·67 cm from the left-hand end.*

2   Find the position of the centroid of the plate shown in Figure 5.8.

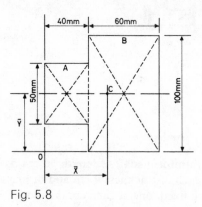

Fig. 5.8

Since the plate is symmetrical about the horizontal axis, the centroid will lie on this line, *i.e.* $\overline{Y} = 50$ mm.

To find $\overline{X}$, take moments about O

$$\text{Moment of whole} = \text{Sum of the moments of the parts}$$

$$\text{Total Area} \times \overline{X} = (\text{Area A} \times 20) + (\text{Area B} \times 70)$$

$$\{(40 \times 50) + (60 \times 100)\}\,\overline{X} = (40 \times 50 \times 20) + (60 \times 100 \times 70)$$

$$8\,000\ \overline{X} = 40\,000 + 420\,000$$

$$\overline{X} = \frac{460 \times 10^3}{8 \times 10^3}$$

$$\overline{X} = 57\cdot5$$

∴ *The centroid of the plate lies 57·5 mm along the axis from the left-hand end.*

3   Figure 5.9 shows an L shaped reinforcing plate. Calculate the position of the centroid of the plate.

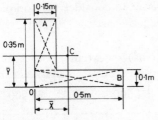

Fig. 5.9

Divide the area as shown into sections A and B.

Part	Area (m²)		x(m)	Ax	y(m)	Ay
A	0·25 × 0·15	0·037 5	0·075	0·002 815	0·225	0·008 45
B	0·5 × 0·1	0·05	0·25	0·012 5	0·05	0·002 5
Totals		0·087 5		0·015 315		0·010 95

$$\overline{X} = \frac{\Sigma Ax}{\Sigma A} \qquad\qquad \overline{Y} = \frac{\Sigma Ay}{\Sigma A}$$

$$= \frac{0·015\,315}{0·087\,5} \qquad\qquad = \frac{0·010\,95}{0·087\,5}$$

$$\overline{X} = 0·175 \qquad\qquad \overline{Y} = 0·225$$

NOTE: C does not lie on the plate.

4   Find the centroid of the plate shown in Figure 5.10.

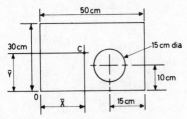

Fig. 5.10

In this case it is better to consider the complete area as positive, and the hole as a negative area.

Part	Area (cm^2)		x(cm)	Ax	y(cm)	Ay
A	30×50	1 500	25	37 500	15	22 500
B	$-\dfrac{\pi}{4} \times 15^2$	-177	35	$-6 190$	10	$-1 770$
Totals		1 323		31 310		20 730

$$\overline{X} = \frac{\Sigma\,Ax}{\Sigma\,A} \qquad\qquad \overline{Y} = \frac{\Sigma\,Ay}{\Sigma\,A}$$

$$= \frac{31\,310}{1\,323} \qquad\qquad = \frac{20\,730}{1\,323}$$

$$\overline{X} = 23{\cdot}6 \qquad\qquad\qquad \overline{Y} = 15{\cdot}6$$

NOTE: There are many instances where it is quicker and easier to subtract one area, rather than split the shape up into a number of parts and then add them all together.

5   Figure 5.11 shows the cross section of a small concrete dam. Determine:
    a) The centroid of the section.
    b) The mass of concrete required if the dam is 10 m long, and concrete has a density of 2 200 kg/m^3

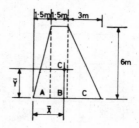

Fig. 5.11

Part	Area (m²)		x(m)	Ax	y(m)	Ay
A	$\frac{1}{2} \times 1\cdot5 \times 6$	4·5	1	4·5	2	9
B	$1\cdot5 \times 6$	9	2·25	20·25	3	27
C	$\frac{1}{2} \times 3 \times 6$	9	4	36	2	18
Totals		22·5		60·75		54

$$\overline{X} = \frac{\Sigma \, Ax}{\Sigma \, A} \qquad \overline{Y} = \frac{\Sigma \, Ay}{\Sigma \, A}$$

$$= \frac{60\cdot75}{22\cdot5} \qquad \frac{54}{22\cdot5}$$

$$\overline{X} = 2\cdot7 \qquad \overline{Y} = 2\cdot4$$

Volume of concrete = Cross sectional area × length
$$= 22\cdot5 \times 10 \text{ m}^3$$
∴ Mass of concrete $= 22\cdot5 \times 10 \times 2\,200$ kg
$$= 495\,000 \text{ kg}$$
∴ *Mass of concrete required = 495 tonnes.*

6   The section of a beam built up from two angle irons and a flat plate
is shown in Figure 5.12(a). For each angle the cross sectional area
is 1 400 mm² and the position of its centroid is indicated in Figure
5.12(b).
Determine the centroid of the beam section, with respect to the
bottom left hand corner.

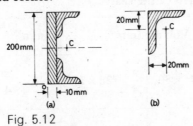

Fig. 5.12

Since the section is symmetrical about the horizontal axis,
$\overline{Y} = 100$ mm.

Part	Area (mm²)		x(mm)	Ax
A	200 × 10	2 000	5	10 000
B		1 400	30	42 000
C		1 400	30	42 000
Total	4 800			94 000

$$\overline{X} = \frac{\Sigma\,Ax}{\Sigma\,A}$$

$$= \frac{94\,000}{4\,800}$$

$$\overline{X} = 19{\cdot}5$$

∴ *Centroid of the section lies 100 mm up from the bottom, and 19·5 mm from the left-hand side.*

7 Find the position of the centre of gravity of the generator shaft shown in Figure 5.13.

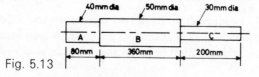

Fig. 5.13

Volumes may be substituted for weights. The centre of gravity will lie on the axis of the shaft.

Part	Volume (mm³)		x (mm)	Vx
A	$\frac{\pi}{4} \times 40^2 \times 80$	$\frac{\pi}{4} \times 128 \times 10^3$	40	$\frac{\pi}{4} \times 5\,120 \times 10^3$
B	$\frac{\pi}{4} \times 50^2 \times 360$	$\frac{\pi}{4} \times 900 \times 10^3$	260	$\frac{\pi}{4} \times 234\,000 \times 10^3$

C	$\dfrac{\pi}{4} \times 30^2 \times 200$	$\dfrac{\pi}{4} \times 180 \times 10^3$	540	$\dfrac{\pi}{4} \times 97\,200 \times 10^3$
Total		$\dfrac{\pi}{4} \times 1\,208 \times 10^3$		$\dfrac{\pi}{4} \times 336\,320 \times 10^3$

$$\overline{X} = \frac{\Sigma\, Vx}{\Sigma\, V} \qquad \overline{X} = \frac{\dfrac{\pi}{4} \times 336\,320 \times 10^3}{\dfrac{\pi}{4} \times 1\,208 \times 10^3}$$

$$\overline{X} = 279$$

∴ *Centre of gravity of the shaft lies 279 mm along the axis from the left-hand end.*

NOTE: By keeping $\pi/4$ as a common factor it eventually cancels out and simplifies the calculation.

8 A plough of mass 500 kg is loaded onto a trailer of mass 400 kg as shown in Figure 5.14.
   a) What is the effect of loading the plough in this position?
   b) Calculate the distance of the centre of gravity from the wheels.
   c) Calculate the force F, in magnitude and direction exerted on the tow bar.

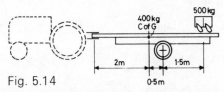

Fig. 5.14

a) *The effect of placing the plough in the position shown is to tip the trailer backwards.* This means that an upward force is exerted on the drawbar, decreasing the reaction on the rear wheels of the tractor and hence decreasing traction.
b) Take moments about the wheel.
   Moments of the whole = Sum of the moments of the parts.
$$900g \times \overline{X} = (500g \times 1\cdot 5) - (400g \times 0\cdot 5)$$

NOTE: Moment of the trailer has a negative sign because it acts in the opposite direction to that of the plough.

98   CENTRE OF GRAVITY

$$i.e. \quad 900\overline{X} = 750 - 200$$

$$\overline{X} = \frac{550}{900}$$

$$= 0.61$$

Hence the centre of gravity of the trailer and the load lies 0·61 m to the rear of the wheels.

c) Take moments about the wheel.

$$F \times 2.5 = 900g \times 0.61 \text{ N}$$

$$F = \frac{900 \times 9.81 \times 0.61}{2.5} \text{ N}$$

$$= 2.16 \text{ kN}$$

The balancing force (*i.e.* that exerted by the tow bar) is 2·16 kN acting downwards. *The force exerted on the towbar is thus 2·16 kN upwards.*

9   A double deck bus has a mass of 8·8 tonne. Its centre of gravity lies 2·08 m above the ground on the central axis and the width of track is 2·44 m.

a) Find the angle to which the bus may be tilted without toppling over.

b) If 20 passengers each of mass 75 kg stand on the upper deck, calculate the angle to which the bus may now be tilted without toppling. Assume the C. of G. of the passengers to be 3.5 m above the ground.

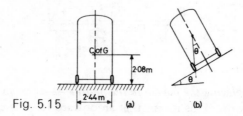

Fig. 5.15

a) $\tan \theta_1 = \dfrac{1.22}{2.08}$

$= 0.585$

$\therefore \quad \theta_1 = 30.3°$

$\therefore$ *Bus can be tilted to 30·3° before toppling.*

b) With passengers on top deck, find new C. of G.

$$\overline{Y}\left(8{\cdot}8+\frac{20\times75}{1\,000}\right) = 8{\cdot}8\times2{\cdot}08+\frac{20\times75}{1\,000}\times3{\cdot}5$$

$$\overline{Y}(8{\cdot}8+1{\cdot}5) = 18{\cdot}3+5{\cdot}25$$

$$\overline{Y} = 2{\cdot}29$$

Hence: $\tan\theta_2 = \dfrac{1{\cdot}22}{2{\cdot}29}$

$$= 0{\cdot}532$$

and $\theta_2 = 28°$

∴ *Bus will now topple at 28°.*

10 The plan and elevation of a small casting are shown in Figure 5.16. Determine the position of the centre of gravity of the casting.

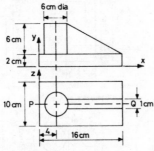

Fig. 5.16

Since the casting is symmetrical about PQ, the C. of G. will lie on this line, *i.e.* $\overline{Z} = 5$ cm.

Part	Volume (cm^3)		x (cm)	Vx	y (cm)	Vy
A	$\frac{\pi}{4}\times6^2\times6$	170	4	680	5	850
B	$\frac{1}{2}\times6\times9\times1$	27	10	270	4	108
C	$16\times10\times2$	320	8	2 560	1	320
Total		517		3 510		1 278

$$\overline{X} = \frac{\Sigma\,Vx}{\Sigma\,V} \qquad\qquad \overline{Y} = \frac{\Sigma Vy}{\Sigma\,V}$$

$$= \frac{3\,510}{517} \qquad\qquad = \frac{1\,278}{517}$$

i.e. $\overline{X} = 6{\cdot}8$ $\qquad\qquad$ *i.e.* $\overline{Y} = 2{\cdot}47$

Examples

1 (Figure 5.17) Find the position of the centroid of each of the shapes (a) to (k).

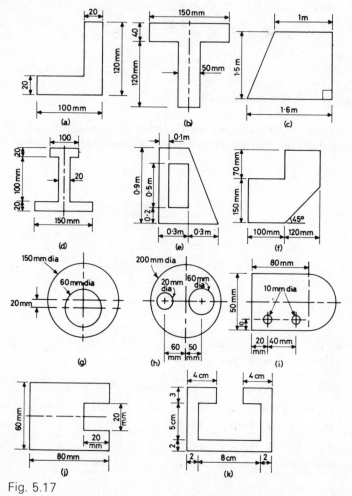

Fig. 5.17

2 (Figure 5.18) A signal lever consists of a flat bar, of mass 5 kg welded to a triangular plate which acts as a counterbalance. The plate has a mass of 2·5 kg. A hole is to be drilled at the centre of gravity of the complete lever. Find the position of this hole with respect to the left hand end of the lever.

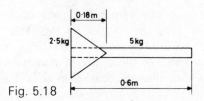

Fig. 5.18

3 (Figure 5.19) The rotor of a small electric motor consists of the armature and the commutator pressed onto a uniform shaft. The masses of the parts are as follows: shaft 0·1 kg, armature 0·5 kg, commutator 0·05 kg. Determine the centre of gravity of the rotor.

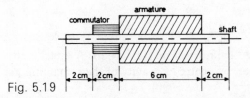

Fig. 5.19

4 Figure 5.20 shows a small casting A, of mass 1·75 kg bolted onto a lathe face plate. A cylindrical counterweight of mass 1·25 kg is attached at B. Find the position of the counterweight which causes the centre of gravity of the assembly to lie on the machine axis.

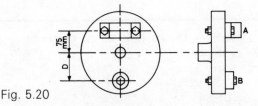

Fig. 5.20

5 (Figure 5.21) A water tank consists of a cylinder with a hemispherical end. Calculate the position of the centre of gravity of the tank when it is full of water.

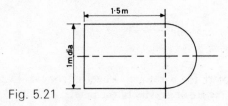

Fig. 5.21

6 (Figure 5.22) One of the rollers (mass 3·2 t) of a small rolling mill is shown at (a). Calculate the position of the centre of gravity of the roller. It has to be lifted by means of two slings, each 1 m long placed 0·8 m on each side of the centre of gravity, and attached to a crane hook (b). Calculate the tension in each sling.

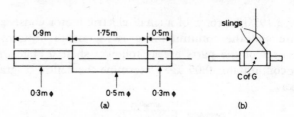

Fig. 5.22

7 (Figure 5.23) A cylindrical bar has part A machined out as shown. Calculate the centre of gravity of the remaining part.

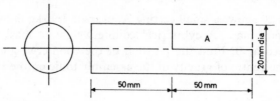

Fig. 5.23

8 (Figure 5.24) Determine the position of the centre of gravity of the crank lever as shown.

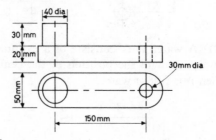

Fig. 5.24

9 (Figure 5.25) A pawl for a ratchet wheel is shown. Determine the position of the centre of gravity of the pawl.

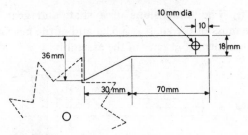

Fig. 5.25

10 (Figure 5.26) A guide rod is supported by a steel lever of thickness 10 mm and spring arrangement. Determine:
a) The position of the centre of gravity of the lever.
b) The tension in the spring when the guide rod exerts no force on the lever. The density of steel is $7·7 \times 10^3$ kg/m³.

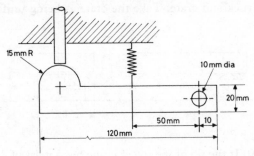

Fig. 5.26

11 (Figure 5.27) The casting has to be lifted by means of a ring bolt attached to it. Find:
a) The position in which the bolt must be for the casting to remain horizontal when lifted.
b) The mass of the casting if the density of cast iron is $7·25 \times 10^3$ kg/m³.

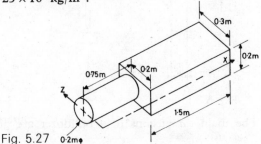

Fig. 5.27 0·2mφ

12 (Figure 5.28) The dimensions of a shaft and gear wheel are
    shown. The mass of the shaft is 3.5 kg and of the gear wheel 2 kg.
    Calculate the centre of gravity of the assembly.

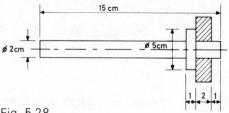

Fig. 5.28

13 (Figure 5.29) The position of the centre of gravity of a truck,
    of mass 3 t, is shown. A crate, of mass 2·5 t, is placed in the given
    position. Calculate the position of the centre of gravity of the
    combined truck and crate. Take the crate as being uniform.

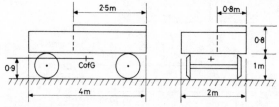

Fig. 5.29

14 (Figure 5.30) If the jib of the tower crane has a mass of 2 t and the
    counterweight has a mass of 2·5 t, determine the greatest mass
    that the crane can lift at the extreme end of the jib in order that
    the centre of gravity will not lie to the right of the central tower
    (*i.e.* to the right of A).

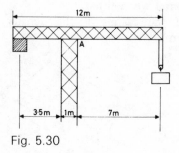

Fig. 5.30

15 (Figure 5.31) The main dimensions of a bulldozer are given.
    Determine:

a) The angle of the maximum slope up which it could travel, without toppling backwards.
b) The angle of the maximum slope across which it could travel, without toppling sideways.

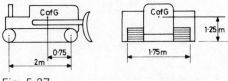

Fig. 5.27

16 (Figure 5.32) Two blocks A and B, of uniform thickness, are placed on top of block C. Is the combination of A and B in stable equilibrium? If not, determine the least distance that block A would need to be moved to make it stable.

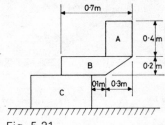

Fig. 5.31

6
Framed Structures

Members   A framework or framed structure (Figure 6.1) consists of an assembly of rods, bars, beams etc. called *Members*. Each member is hinged or connected at its ends to other parts of the framework structure.

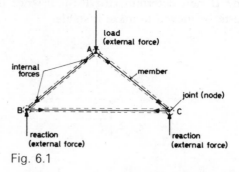

Fig. 6.1

The External Forces   act at the joints only, and for design purposes these joints (called nodes) are assumed 'pin jointed' and frictionless.

The Internal Forces   (forces within each member resisting external forces) are therefore direct tensile or compressive forces (Figure 6.2).

The 'pin joint' can be taken as the point of application of each force in a single member and the magnitude of this internal force can be determined by the triangle or polygon of forces. These forces are assumed to act at one point of the joint, but in practice the joints are riveted, welded or otherwise connected; therefore the calculated values of tension and compression in the members will be on the safe side. The magnitude of the internal forces is used in conjunction with the allowable stress for the material to determine the cross-sectional area of each member.

Nature of Internal Forces Figure 6.2a shows a member AB in *Tension*. The external forces F_1 act outwards tending to stretch the member. In Figure 6.2b the member CD is in *Compression*. The external forces F_2 act inwards tending to crush the member.

Each case illustrates that the internal forces resist the action of external forces and the direction of the arrowheads in the members determine the *nature* of the forces. In Figure 6.1 it can be seen that members AB and AC are in compression (struts) and member BC is in tension (tie).

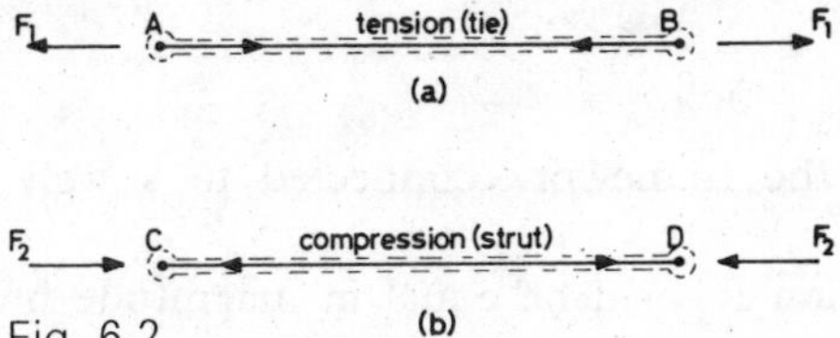

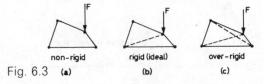

(a)

(b)

Fig. 6.2

An ideal frame (rigid) is one which has just sufficient members to prevent it changing shape or collapsing under any system of loading (Figure 6.3).

non-rigid   rigid (ideal)   over-rigid

Fig. 6.3 (a)   (b)   (c)

Figure 6.3a: The shape can change under the given force F. The structure is *non-rigid*.

Figure 6.3b: The diagonal member prevents a change of shape. The structure is *rigid*.

Figure 6.3c: The extra diagonal member is unnecessary for static loading. The structure is *over-rigid*.

In a simple ideal frame the whole framework or structure consists of a number of triangles. The external force system must be in equilibrium and each joint in the structure must also be in equilibrium under the system of concurrent forces.

Wind Forces The components F_1 and F_2 of wind forces act perpendicular to the face of the framework or truss as shown in Figure 6.4.

Reactions To allow for temperature change (expansion and contraction) framed structures such as bridges etc. rest on rollers (Figure 6.4). For reactions at rollers see page 8.

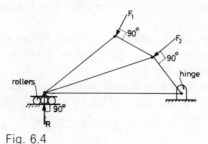

Fig. 6.4

Consider the framework connected to a wall as shown in Figure 6.5.

The reaction R_1 will be equal in magnitude but opposite in direction to the internal force in AB.

The reaction R_2 will be the equilibrant of the internal forces in members CB and CD. The direction of R_2 can be found by considering the equilibrium of the whole external force system.

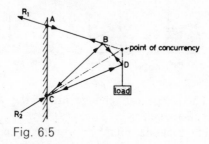

Fig. 6.5

Method for graphical solution

1 Draw the Space diagram to a suitable scale and show all external forces, reactions etc., acting on the framework.
2 Using Bow's notation (see page 12)
   a) letter the spaces between the external forces.
   b) letter the spaces between the members.
3 Calculate any external forces if necessary.
4 Draw the Vector (Force) diagram to a suitable scale starting with a known external force. The complete external force polygon may be a straight line.

5 Complete the Vector diagram by drawing the triangle or polygon of forces for each joint superimposed on one another.
6 Tabulate results, or mark them on each member in the Space diagram. Indicate the direction of the reactions on the Space diagram.

The Space and Vector diagrams should be kept as close together as possible to facilitate easy use of drawing instruments. No arrowheads should be shown on the Vector diagram. A high degree of accuracy is required; lines should be thin and measurements exact.

Worked Examples

1 Find the reactions at the supports and the magnitude and nature of the forces in the members of the simple roof truss shown.

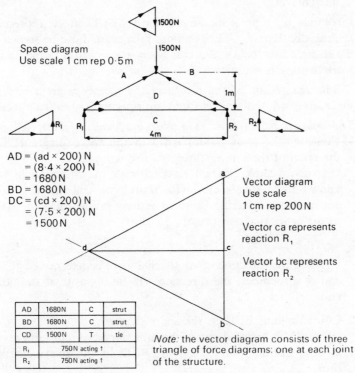

Space diagram
Use scale 1 cm rep 0·5 m

$AD = (ad \times 200)$ N
$= (8·4 \times 200)$ N
$= 1680$ N
$BD = 1680$ N
$DC = (cd \times 200)$ N
$= (7·5 \times 200)$ N
$= 1500$ N

Vector diagram
Use scale
1 cm rep 200 N

Vector ca represents reaction R_1

Vector bc represents reaction R_2

AD	1680N	C	strut
BD	1680N	C	strut
CD	1500N	T	tie
R_1	750N acting ↑		
R_2	750N acting ↑		

Note: the vector diagram consists of three triangle of force diagrams: one at each joint of the structure.

Fig. 6.6

Difficulty often arises in:

a) drawing the vector diagram.

b) determining the nature of the forces in the members (transferring directions from the vector diagram to the space diagram).

To draw the vector diagram :

1 Draw 'ab' parallel to the known external force AB (1500 N) to the given scale.

2 Through 'a' draw a straight line parallel to AD. (Note that the line extends on both sides of 'a' and the point 'd' lies somewhere on this line.)

3 Through 'b' draw a line parallel to BD (point 'd' lies somewhere on this line). The exact position of 'd' is the intersection of the lines drawn through 'a' and 'b'.

4 Through 'd' draw a line parallel to CD. Point 'c' is where this line cuts 'ab'.

NOTE that 'bc' represents the reaction R_2(BC) and 'ca' represents the reaction R_1(CA). The reactions if calculated can be shown on the straight line 'ba' (in this case $R_1 = R_2 = 750\,N$ as the framework loading is symmetrical).

The magnitude of the forces in the members are found by measuring 'ad' 'bd' and 'dc' and multiplying by the scale factor.

The nature of the forces in the members are found as follows :

1 Consider the joint (node) ABD in the space diagram. The direction of the known force AB is downwards.

    In the vector diagram starting with the direction of this known force follow clockwise round the triangle ab↓ bd↖ and da↗ transferring these directions round the joint ABD in the space diagram thus:

    When the direction of an internal force is determined at one end of a member, the direction will be opposite at the other end.

2 Consider joint BCD in the space diagram.

    From the vector diagram the directions bc↑ cd← and db↘ are transferred round the joint BCD in the space diagram thus:

3 Consider joint CAD in the space diagram.

From the vector diagram the directions ca↑ ad ↙ and dc→ are transferred round the joint CAD in the space diagram thus: A↙D ↑C

The magnitude and nature of the forces are now determined and tabulated as shown in Figure 6.6.

NOTE: In this example the magnitude of the reactions can be found direct from the vector diagram. In later examples it perhaps will be necessary to calculate the reactions (by application of moments) before the vector diagram can be completed.

— 7   Figures 6.7 to 6.12 illustrate the method of solving simple frame problems.

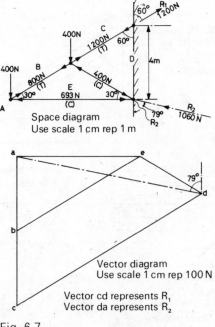

Space diagram
Use scale 1 cm rep 1 m

Vector diagram
Use scale 1 cm rep 100 N

Vector cd represents R₁
Vector da represents R₂

Fig. 6.7

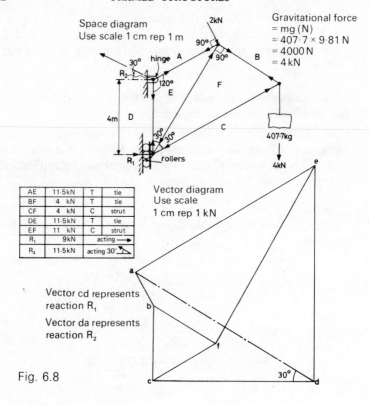

AE	11·5kN	T	tie
BF	4 kN	T	tie
CF	4 kN	C	strut
DE	11·5kN	T	tie
EF	11 kN	C	strut
R₁	9kN	acting ⟶	
R₂	11·5kN	acting 30°	

R_1

R_2

Vector diagram
Use scale
1 cm rep 1 kN

Vector cd represents
reaction R_1

Vector da represents
reaction R_2

Fig. 6.8

Gravitational force
= mg (N)
= 407·7 × 9·81 N
= 4000 N
= 4 kN

407·7kg

4kN

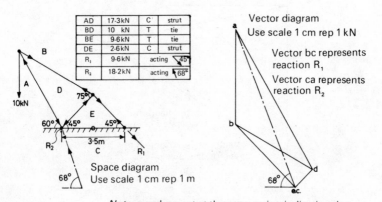

AD	17·3kN	C	strut
BD	10 kN	T	tie
BE	9·6kN	T	tie
DE	2·6kN	C	strut
R₁	9·6kN	acting 45°	
R₂	18·2kN	acting 68°	

Vector diagram
Use scale 1 cm rep 1 kN

Vector bc represents
reaction R_1

Vector ca represents
reaction R_2

Space diagram
Use scale 1 cm rep 1 m

Fig. 6.9

Note: e and c meet at the same point, indicating that member EC carries no load and is termed redundant. In this case EC is the ground.

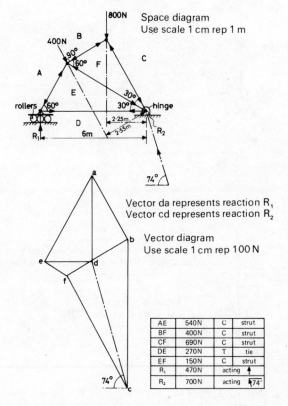

Space diagram
Use scale 1 cm rep 1 m

Vector da represents reaction R_1
Vector cd represents reaction R_2

Vector diagram
Use scale 1 cm rep 100 N

AE	540N	C	strut
BF	400N	C	strut
CF	690N	C	strut
DE	270N	T	tie
EF	150N	C	strut
R_1	470N	acting ↑	
R_2	700N	acting ↖74°	

R_1 can also be found by calculation:
Take moments about hinge
Σ C.M. = Σ A.C.M.
$$6R_1 = (400 \times 2.55) + (800 \times 2.25)\text{N}$$
$$R_2 = \frac{2820}{6} \text{ N}$$
$$R_1 = 470\text{N}$$

Check: da = 470N

Fig. 6.10

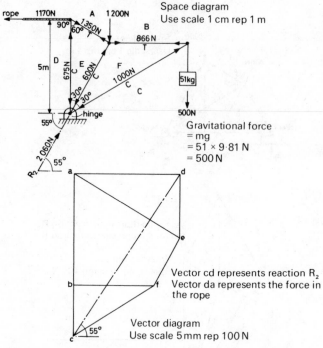

Fig. 6.11

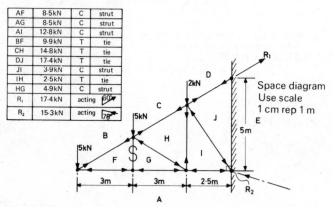

AF	8·5kN	C	strut
AG	8·5kN	C	strut
AI	12·8kN	C	strut
BF	9·9kN	T	tie
CH	14·8kN	T	tie
DJ	17·4kN	T	tie
JI	3·9kN	C	strut
IH	2·5kN	T	tie
HG	4·9kN	C	strut
R_1	17·4kN	acting	60°
R_2	15·3kN	acting	78°

Fig. 6.12
(continued on next page)

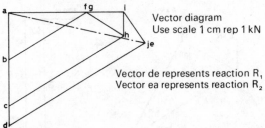

Vector diagram
Use scale 1 cm rep 1 kN

Vector de represents reaction R₁
Vector ea represents reaction R₂

Fig. 6.12
(continued)

Note: f and g meet at the same point. This indicates that member FG carries no load and is redundant; similarly with j and e – in this case JE is the wall.

Examples

Using the methods shown in Figure 6.6 determine in Examples 1 to 12 below:

a) The magnitude and nature of the forces in the members.

b) The magnitude and direction of the reactions.

Tabulate results or mark the forces on the members in the space diagram.

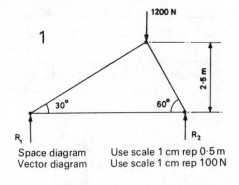

1

1200 N

30° 60°

2·5 m

R₁ R₂

Space diagram   Use scale 1 cm rep 0·5 m
Vector diagram   Use scale 1 cm rep 100 N

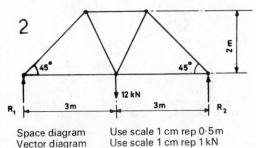

2

45° 45°

2 m

12 kN

R₁ 3 m 3 m R₂

Space diagram   Use scale 1 cm rep 0·5 m
Vector diagram   Use scale 1 cm rep 1 kN

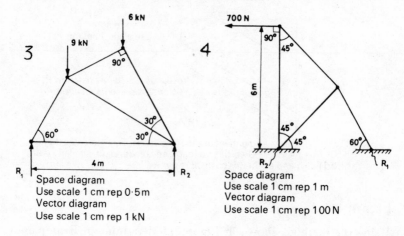

3

Space diagram
Use scale 1 cm rep 0·5 m
Vector diagram
Use scale 1 cm rep 1 kN

4

Space diagram
Use scale 1 cm rep 1 m
Vector diagram
Use scale 1 cm rep 100 N

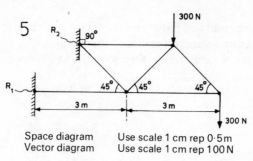

5

Space diagram   Use scale 1 cm rep 0·5 m
Vector diagram   Use scale 1 cm rep 100 N

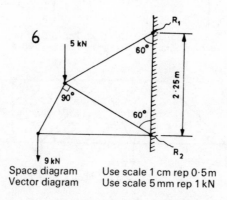

6

Space diagram   Use scale 1 cm rep 0·5 m
Vector diagram   Use scale 5 mm rep 1 kN

7

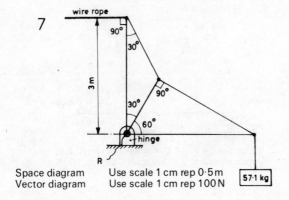

Space diagram   Use scale 1 cm rep 0·5 m
Vector diagram   Use scale 1 cm rep 100 N

8

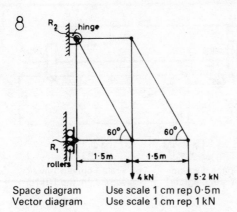

Space diagram   Use scale 1 cm rep 0·5 m
Vector diagram   Use scale 1 cm rep 1 kN

9

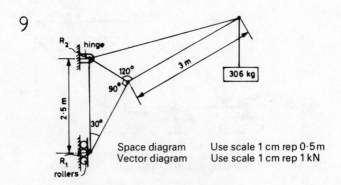

Space diagram   Use scale 1 cm rep 0·5 m
Vector diagram   Use scale 1 cm rep 1 kN

10

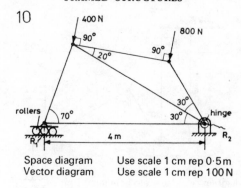

Space diagram   Use scale 1 cm rep 0·5 m
Vector diagram   Use scale 1 cm rep 100 N

11

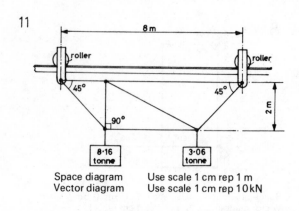

Space diagram   Use scale 1 cm rep 1 m
Vector diagram   Use scale 1 cm rep 10 kN

12

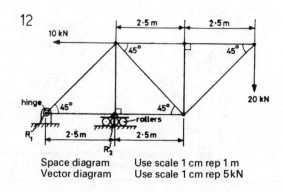

Space diagram   Use scale 1 cm rep 1 m
Vector diagram   Use scale 1 cm rep 5 kN

7
Friction

Whenever one object slides, or tends to slide over another, a resisting force is brought into play between the surfaces in contact. The force is called the *Force of Friction* and always acts in a direction opposite to that of the relative motion.

Experimental laws have been discovered, and these form the basis from which friction problems are tackled.

The Laws of Dry Friction

1 Frictional resistance is proportional to the thrust between the surfaces.
2 Frictional resistance depends on the nature (materials) and condition (roughness) of the surfaces in contact.
3 Frictional resistance is independent of the area of the surfaces in contact.
4 Frictional resistance is independent of the speed of sliding (within normal limits).

The Coefficient of Friction   From law 1 it can be seen that:
$F \propto R_N$ where F is the frictional force created and R_N (the normal reaction) is the thrust between the surfaces.

thus $F = \text{constant} \times R_N$

i.e. $F = \mu \times R_N$ where μ (the Greek letter mu) is the name given to the constant.

or $\mu = \dfrac{F}{R_N}$

μ is known as the coefficient of friction.

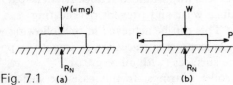

Fig. 7.1   (a)   (b)

If we consider a block sitting on a flat surface (Figure 7.1a) it is in equilibrium and thus the normal reaction R_N is equal to the weight W of the block. If a force P is gradually applied (Figure 7.1b) it will try to move the block, and a resisting frictional force F will be created between the surfaces, equal in magnitude to P. As P is increased it reaches a value where the block is about to move, and immediately it moves, the force required to keep it moving at a steady speed falls to a lower value.

The value of friction F_S just on the point of motion is known as the limiting value of *Static Friction*, and the corresponding value of the coefficient is known as the coefficient of static friction: *i.e.* $F_S = \mu_S R_N$

When the block is moving the friction is known as *Kinetic Friction* (F_K) and: $F_K = \mu_K R_N$

In all cases the static friction is greater than the kinetic friction. This may be simply illustrated as in Figure 7.2. The actual surfaces though seeming smooth have slight surface irregularities which can be seen only with the aid of a magnifying glass. When the block is stationary, as in (a), the 'hills' of one surface inter-lock with the 'hollows' of the other, and more force is required to lift the block out of the hollows. When the block is moving, however, (b), it never completely settles into the hollows and thus a smaller force is required.

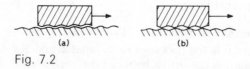

(a)   (b)

Fig. 7.2

Friction is not always undesirable and there are situations where it is important to have as much friction as possible: *e.g.* between the wheels of a car and the ground; in friction clutches; in brakes. Special materials have been developed for these purposes which have high friction properties.

In other cases it is important to reduce friction to a minimum, in order to reduce wear and prevent overheating: *e.g.* bearings; machine slides. This can be achieved in the following ways:
1 Ensure that the contact surfaces are as smooth as possible.
2 Lubricate the surfaces (with oil or grease).
3 Use ball or roller bearings. In this case the surfaces roll over

each other rather than sliding and the force required is very much reduced.

4 Use special materials. Plastics such as nylon and PTFE can often be used instead of conventional materials.

NOTE: In every case $\mu < 1$. Some representative values of μ are given in the appendix.

Inclined Forces

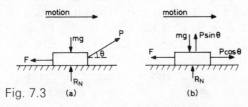

Fig. 7.3 (a) (b)

If the applied force P acts at an angle (Figure 7.3a), then it tends to lift the block off the surface. The method of solution is to resolve P into rectangular components (Figure 7.3b). Since the block is in equilibrium:

$$F = P \cos \theta \quad \text{and} \quad R_N = mg - P \sin \theta$$

$$\text{hence } \mu = \frac{F}{R_N} = \frac{P \cos \theta}{mg - P \sin \theta}$$

Inclined Plane

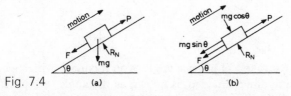

Fig. 7.4 (a) (b)

The method, as before, is to resolve all forces parallel and perpendicular to the plane surface.

$$F = P - mg \sin \theta \quad \text{and} \quad R_N = mg \cos \theta$$

$$\mu = \frac{F}{R_N} = \frac{P - mg \sin \theta}{mg \cos \theta}$$

It is not recommended that these equations be memorised but that each problem be solved from basic principles, as shown in the worked examples.

Angle of Friction

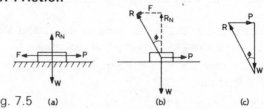

Fig. 7.5   (a)   (b)   (c)

The block (Figure 7.5a) has two external forces P and W and two reactions F and R_N. If the resultant reaction (R) of F and R_N is found, then we have three forces as in Figure 7.5b and the force triangle can be drawn, or Lami's Theorem used to find the unknown force.

The angle between R and R_N is known as the *Angle of Friction* and is given the symbol ϕ (phi). The greater the friction component the greater the angle ϕ.

NOTE:
$$\tan \phi = \frac{F}{R_N} = \mu$$

Figure 7.6 shows the principle applied to an inclined plane with the force P at an angle α to the plane.

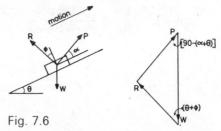

Fig. 7.6

Angle of Repose   The least angle of an incline at which a body will just slide down unaided, is known as the angle of repose. Its value is equal to the angle of friction for the given surfaces.

Friction Torque   With a rotating system such as a shaft running in bearings, the frictional force produces a torque opposing the rotation.

$$F = \mu R_N$$
$$= \mu \, mg$$
$$\text{torque} = \text{force} \times \text{radius}$$
$$T_F = (\mu \, mg) \, r$$

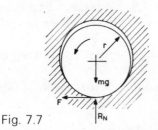

Fig. 7.7

Worked Examples

1 A wooden box of mass 45 kg rests on a wooden floor (Figure 7.8).
   If it requires a force of 220 N to move the box, calculate the
   limiting value of the coefficient of friction (coefficient of static
   friction). It then requires a force of 176 N to keep it moving
   after it has started. Find the coefficient of kinetic friction.

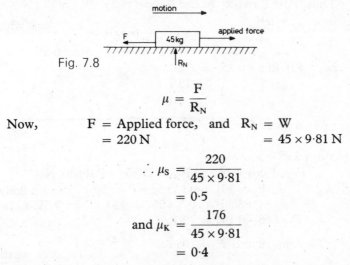

Fig. 7.8

$$\mu = \frac{F}{R_N}$$

Now,   F = Applied force, and $R_N = W$
        = 220 N $= 45 \times 9\cdot81$ N

$$\therefore \mu_S = \frac{220}{45 \times 9\cdot81}$$

$$= 0\cdot5$$

$$\text{and } \mu_K = \frac{176}{45 \times 9\cdot81}$$

$$= 0\cdot4$$

2 If the box in question 1 has 55 kg of scrap metal placed in it, find
   the horizontal force now required to keep it moving steadily.

   F = Applied force, $R_N = (45 + 55)9\cdot81$ N

$$F = \mu R_N$$
$$F = 0\cdot4 \times 100 \times 9\cdot81 \text{ N}$$

   i.e. Required force = 392 N

3 A steel block (Figure 7.9a) of mass 25·5 kg rests on a horizontal surface. The coefficient of kinetic friction between the surfaces is 0·3. Calculate:

a) The value of the force P required to keep the block moving steadily.

b) The value of P required if it is reversed as shown in Figure 7.9b.

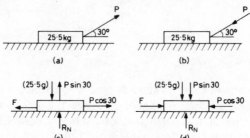

Fig. 7.9

a) Resolve P horizontally and vertically.

Thus, $F = P \cos 30° \, N$ and $R_N = (25\cdot5g - P \sin 30°) \, N$

But, $F = \mu R_N$

$$P \cos 30° = \mu (25\cdot5g - P \sin 30°)$$

i.e. $0\cdot866P = 0\cdot3 \times 25\cdot5 \times 9\cdot81 - 0\cdot3 \times 0\cdot5P$

i.e. $P(0\cdot866 + 0\cdot15) = 75\cdot1$

$$P = \frac{75\cdot1}{1\cdot016}$$

∴ *The required force* $P = 74 \, N$

b) In this case:

$$F = P \cos 30 \, N \text{ and } R_N = (25\cdot5g + P \sin 30) \, N$$
$$P \cos 30° = \mu (25\cdot5g + P \sin 30)$$

i.e. $0\cdot866P = 0\cdot3 \times 25\cdot5 \times 9\cdot81 + 0\cdot3 \times 0\cdot5P$

i.e. $P(0\cdot866 - 0\cdot15) = 75\cdot1$

∴ *The required force* $P = 105 \, N$

4 A block A of mass 20 kg rests on top of block B of mass 15 kg (Figure 7.10). If μ between A and B is 0·5 and μ between B and C is 0·2, calculate:

a) The force P required to make the block A slide over the block B.

b) The corresponding tension T in the rope.

c) The blocks are reversed as shown in (b).
Calculate the force P now required to move block B to the left

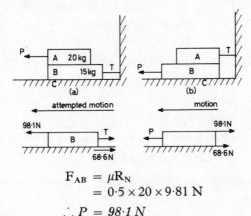

Fig. 7.10

a)
$$F_{AB} = \mu R_N$$
$$= 0.5 \times 20 \times 9.81 \text{ N}$$
$$\therefore P = 98.1 \text{ N}$$

b) This force is tending to pull block B forward.

$$F_{BC} = \mu R_N$$
$$= 0.2 \times 35 \times 9.81 \text{ N}$$
$$= 68.6 \text{ N}$$

For block B:

$$T + 68.6 \text{ N} = 98.1 \text{ N}$$
$$\therefore \textit{Tension } T = 29.5 \text{ N}$$

c) In this case:

$$P = 98.1 \text{ N} + 68.6 \text{ N}$$

$$\therefore \textit{Force to pull block B forward is 166.7 N.}$$

5 A machine guide block of mass 2·04 kg runs in horizontal slides
(Figure 7.11). The coefficient of friction between the surfaces is
0·15 and the connecting rod exerts a force of 80 N. Calculate:
a) The frictional force opposing motion with the connecting rod
in the position shown.
b) The angle of the connecting rod for which there will be no
frictional force acting.

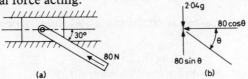

Fig. 7.11   (a)   (b)

a) Resolving the applied force vertically and horizontally, 80 sin 30° pushes the block upwards,

$$\therefore R_N = (2{\cdot}04 \times 9{\cdot}81 - 80 \sin 30°)\,N$$
$$R_N = 20\,N - 40\,N$$
$$R_N = -20\,N$$

i.e. the rod is pushing the block upwards against the top guide.

$$F = \mu R_N$$
$$= 0{\cdot}15 \times 20\,N$$

i.e. Frictional force = 3 N

b) For no friction force the normal reaction must be zero,

$$\therefore 2{\cdot}04\,g = 80 \sin \theta$$
$$\sin \theta = \frac{2{\cdot}04 \times 9{\cdot}81}{80} = 0{\cdot}25$$
$$\therefore \theta = 14{\cdot}5°$$

6   A block of mass 10 kg slides at a uniform rate down a plane inclined at 25° to the horizontal (Figure 7.12). Determine:

a) The coefficient of friction between the surfaces.

b) The force P, parallel to the plane to make it move up the plane at a uniform speed.

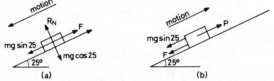

Fig. 7.12   (a)   (b)

a) Resolving the gravity force gives:

$$R_N = mg \cos 25° \quad \text{and} \quad F = mg \sin 25°$$
$$\mu = \frac{F}{R_N} = \frac{mg \sin 25°}{mg \cos 25°}$$
$$= \tan 25°$$
$$\therefore \mu = 0{\cdot}465$$

b) In this case motion is up the plane and thus friction acts down (Figure 7.12b)

$$P = F + mg \sin 25°$$
$$= \mu(mg \cos 25°) + mg \sin 25°$$

$$= [0.46 (10 \times 9.81 \times 0.91) + 10 \times 9.81 \times 0.42 \ N]$$
$$= 41.1 \,N + 41.4 \,N$$

$\therefore$ *Force required* = *82.5 N*

7   A lathe is slid from a lorry down a ramp inclined at 35° to the horizontal (Figure 7.13). It has a mass of 160 kg and is supported by a horizontal rope attached to a winch. Calculate the tension in the rope when the machine slides down the ramp at a steady speed. (μ between machine base and ramp is 0·45.)

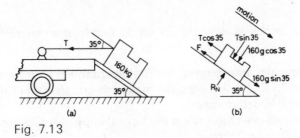

Fig. 7.13

Resolving all forces (Figure 7.13b):

$$F + T \cos 35° = mg \sin 35° \quad \text{and} \quad R_N = T \sin 35° + mg \cos 35°$$
$$\text{i.e. } \mu(T \sin 35° + mg \cos 35°) + T \cos 35° = mg \sin 35°$$
$$0.45(T \times 0.575 + 160 \times 9.81 \times 0.82) + T \times 0.82 = 160 \times 9.81 \times 0.575$$
$$0.259\,T + 580 + 0.82\,T = 904$$
$$T = \frac{(904 - 580)}{(0.82 + 0.259)} = \frac{324}{1.079}$$

Tension in the rope = *301 N*

8   If in Example 9, Chapter 2, the plane is not smooth, and the coefficient of friction between the surfaces is 0·4, calculate the tension T in the rope to move the box up the plane at uniform speed.

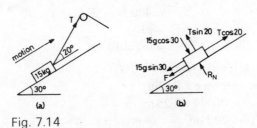

Fig. 7.14

$$F = (T\cos 20° - 15\,g\sin 30°)\,N \text{ and } R_N = (15\,g\cos 30° - T\sin 20°)\,N$$

$$\mu = \frac{F}{R_N} = \frac{T\cos 20° - 15\,g\sin 30°}{15\,g\cos 30° - T\sin 20°}$$

$$i.e. \quad 0\cdot 4 = \frac{0\cdot 94\,T - 15\times 9\cdot 81\times 0\cdot 5}{15\times 9\cdot 81\times 0\cdot 866 - 0\cdot 34\,T}$$

$$51 - 0\cdot 136\,T = 0\cdot 94\,T - 73\cdot 6$$

$$T = \frac{124\cdot 6}{1\cdot 076}$$

$$= 116$$

∴ *The tension to move the box up the incline at a uniform rate is 116 N.*

9 The braking arrangement for a slow running shaft is shown (Figure 7.15). If μ between the brake pad and the rim of the fly-wheel is 0·7, determine:

a) The frictional force when the force P applied at C is 100 N.

b) The frictional torque.

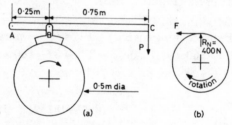

Fig. 7.15   (a)   (b)

a) Take moments about A

$$\Sigma \text{ C.M.} = \Sigma \text{ A.C.M.}$$

thus $(100\times 1) = R_B\times 0\cdot 25$

$$R_B = \frac{100}{0\cdot 25}\,N$$

$$= 400\,N = R_N$$

$$F = \mu R_N$$

$$= 0\cdot 7\times 400\,N$$

∴ *Frictional force created is 280 N.*

b) Torque = Force × Radius

i.e. $T_F = 280\,N\times 0\cdot 25\,m$

$$T_F = 70\,Nm$$

10 A uniform shaft of diameter 25 mm runs in bronze bearings at each end and has a drum of diameter 100 mm placed centrally on it (Figure 7.16). The mass of the shaft and drum is 24 kg. It is found that a mass of 0·45 kg attached to a string wound round the drum is just sufficient to rotate the shaft at a constant speed. Calculate the coefficient of friction between the shaft and the bearing.

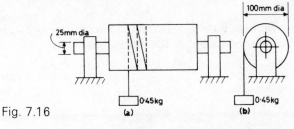

Fig. 7.16

$$\text{Applied torque } (T_A) = (0.45 \times 9.81)50 \text{ Nmm}$$
$$T_A = \text{Frictional torque } (T_F)$$
$$i.e. \quad T_F = 221 \text{ Nmm}$$
$$T_F = F \times r$$
$$F = \frac{T_F}{r} = \frac{221}{12.5} \text{ N} = 17.7 \text{ N}$$
$$\mu = \frac{F}{R_N} = \frac{17.7}{24 \times 9.81}$$
$$\therefore \quad \mu = 0.075$$

11 A grab is used in a quarry for lifting blocks of stone (Figure 7.17). For the grab shown, calculate the minimum value of the coefficient of friction to enable the blocks to be lifted.

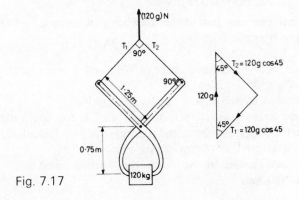

Fig. 7.17

Tension in each rope is 120g cos 45°

Take moments about O

$$\Sigma \text{ C.M.} = \Sigma \text{ A.C.M.}$$
$$120 \text{ g cos } 45° \times 1·25 = R_N \times 0·75$$
$$R_N = \frac{120\text{g cos } 45° \times 1·25}{0·75} \text{ N}$$
$$= 1390 \text{ N}$$

There are two rubbing surfaces; ∴ total frictional force = $2 \times \mu R_N$. This must balance the weight of the block.

$$\therefore \quad 2 \times \mu \times 1\,390 = 120 \times 9·81$$
$$\mu = \frac{120 \times 9·81}{1\,390 \times 2}$$
$$i.e. \quad \mu = 0·424$$

12 An engine draws 8 wagons, each of mass 9 t along a level track. The tractive resistance is 66 N/t. Determine:

   a) The drawbar pull.

   b) The force in the coupling between the fourth and fifth wagons.

   a)   Force = Tractive resistance × mass
                                        = $66 \times (8 \times 9)$ N
                                        = 4 750 N

   ∴ *Drawbar pull* = *4·75 kN*

   b) There are four wagons after the coupling

   $$\therefore \quad F = \text{T.R.} \times (4 \times 9)$$
   $$= 66 \times 36 \text{ N}$$

   ∴ *Force in the coupling is 2·375 kN*

   i.e. in this particular case half of the drawbar pull because half of the wagons are after the coupling.

13 A deltic diesel locomotive has a mass of 108 t and μ between the wheels and rails is 0·12. Calculate:

   a) The maximum drawbar pull of the locomotive.
       The engine draws trucks, each of mass 12 t. If the frictional resistance for both locomotive and trucks is 250 N/t, Calculate:

   b) The maximum number of trucks that may be pulled.

   c) The total force exerted by the locomotive when coupled to this number of trucks.

a) $$F = \mu R_N = 0.12 \times 108 \times 9.81 \text{ kN}$$
$$= 127 \text{ kN}$$

$\therefore$ *Maximum drawbar pull = 127 kN*

b) Force to move 1 truck $= \text{T.R.} \times \text{Mass}$
$$= 250 \times 12 \text{ N}$$
$$= 3 \text{ kN}$$

$\therefore$ Number of trucks $= \dfrac{127}{3} = 43 \cdot 3$

Hence maximum number of trucks is 43

c) Total force $=$ Force to move the trucks $+$ force to move engine
$$= (43(250 \times 12) + 250 \times 108) \text{ kN}$$
$$= 156 \text{ kN}$$

$\therefore$ *Force exerted by the engine is 156 kN.*

14 A car of mass 850 kg tows a caravan of mass 750 kg up an incline of 1 in 8 (Figure 7.18). The tractive resistance of the caravan is 450 N/1 000 kg. Calculate:

a) The drawbar pull.

b) The minimum value of μ between the car tyres and the road surface to prevent the wheels from spinning. 60% of the mass of the car may be taken as being above the driving wheels.

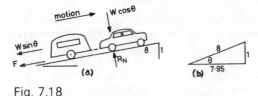

Fig. 7.18

a) $F = \text{T.R.} \times \text{Mass} = 450 \times \dfrac{750}{1\,000} \text{ N} = 338 \text{ N}$

$W \sin \theta = 750 \times 9.81 \times \frac{1}{8} \text{ N} = 920 \text{ N}$

$\therefore$ *Drawbar pull* $= 338 \text{ N} + 920 \text{ N}$
$$= 1\,258 \text{ N}$$

b) $\cos \theta = \dfrac{7.95}{8}$

$R_N = W \cos \theta \text{ N} = \dfrac{60}{100} \times 850 \times 9.81 \times \dfrac{7.95}{8} \text{ N}$

$R_N = 4\,960 \text{ N}$

$$\mu = \frac{F}{R_N} = \frac{1\ 258}{4\ 960}$$
$$\therefore \quad \mu = 0.254$$

Examples

1 A box of mass 12 kg requires a horizontal force of 55 N to keep it moving along a floor at a uniform speed. Find the coefficient of friction between the box and the floor.

2 The coefficient of friction between a bench top and a tool box is 0·55. Calculate the horizontal force required to slide the box along the bench, if the mass of the box is 7·5 kg.

3 A sledge for shifting bales of hay over rough ground has a mass of 125 kg. Calculate:
   a) The coefficient of friction between the sledge and the ground when it requires a force of 870 N to move the sledge.
   b) The force required to move the sledge when loaded with 3 bales, each of mass 45 kg.

4 Find the force required to move a crate, of mass 36 kg, across a floor, when the coefficient of friction between the floor and the crate is 0·4.

5 A lathe tailstock has a mass of 2·55 kg and requires a force of 3·75 N to slide it along the bed of the lathe. Determine the value of the coefficient of friction between the tailstock and the lathe bed.

6 A log, of mass 357 kg, is pulled on to the table of a sawmill. If μ between the log and the table is 0·3, determine the force required.

7 A stone block resting on a horizontal surface has a rope attached to it. The rope is inclined at 45° to the horizontal, and the tension in the rope is 240 N. Determine the minimum mass of the block to prevent it from sliding ($\mu = 0.3$).

8 In a factory assembly line, small rectangular blocks are fed from a hopper onto a conveyor belt by means of an inclined chute (Figure 7.19). It requires a force of 3 N to slide a block, of mass 1·5 kg, along a horizontal piece of the material of which the chute is constructed. Find the minimum angle ϕ of the chute to enable the blocks to slide down on to the conveyor.

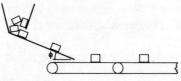

Fig. 7.19

9 A casting of mass 125·5 kg is supported by a rope running over a pulley (Figure 7.20). The rope is anchored to a block resting on a horizontal surface and the limiting value of the coefficient of friction between the surfaces is 0·3. Calculate the least mass of the block required in order to support the casting in the position shown.

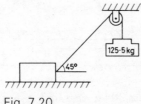

Fig. 7.20

10 A skip sliding on rails is used to load aggregate into a large concrete mixer (Figure 7.21). The mass of the skip is 40 kg and the coefficient of friction between skip and rails is 0·25. Determine:
   a) The force required to move the skip up the rails when it contains 75 kg of aggregate.
   b) The maximum mass that the skip can carry if the tension T in the rope is not to exceed 2 kN.

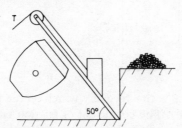

Fig. 7.21

11 A machine of mass 280·5 kg is slid up a loading ramp inclined at 30° to the horizontal, by means of a rope. The coefficient of friction between the ramp and the machine is 0·5. Determine:
   a) The tension in the rope when it is parallel to the ramp.
   b) The tension in the rope if it is horizontal.

12 A steel ingot of mass 1 500 kg has to be slid off the back of a tipping
    lorry (Figure 7.22). If μ between the floor of the lorry and the ingot
    is 0·5, find:
    a) The angle to which the floor must be raised in order that the
        ingot will slide off unaided.
    b) The force exerted by the hydraulic ram just as the ingot is on
        the point of moving.
    c) If the body can only be raised to 20°, what horizontal force P
        will be required to slide the ingot off?

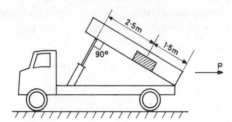

Fig. .22

13 A sliding door of mass 30 kg runs on a nylon track (Figure 7.23).
    Calculate:
    a) The coefficient of friction, if it requires a force of 25 N in order
        to close the door when the track is horizontal.
    b) The angle to the horizontal that the track must be placed in
        order that the door will close itself.

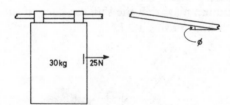

Fig. 7.23

14 A breakdown truck with a recovery winch on the front has a mass
    of 3 t (Figure 7.24). The coefficient of friction between the tyres
    and the road surface is 0·75. It is used to pull a car, of mass 800 kg,
    out of a ditch. Find:
    a) The force required to pull the car out, if μ between the car and
        the ground is 0·8.
    b) The maximum force that can be exerted by the winch with the
        cable at the same angle.

FRICTION

135

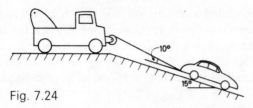

Fig. 7.24

15 The tension in a rope running round a 150 mm diameter pulley is
1·2 kN (Figure 7.25). The spindle diameter is 20 mm and the
coefficient of friction between the spindle and bearing is 0·08.
Determine:
a) The frictional force created at the spindle.
b) The frictional torque.

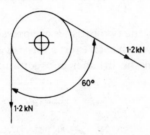

Fig. 7.25

16 A piece of wood is pressed against a sanding disc with a force of
45 N (Figure 7.26). The coefficient of friction between the disc
and the wood is 0·9. Calculate:
a) The frictional force created.
b) The frictional torque.
c) The power absorbed in friction if the disc runs at 850 rev/min.

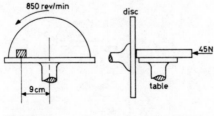

Fig. 7.26

17   A water jet hitting a turbine rotor is shown in the sketch, and is just sufficient to rotate it at constant speed when the turbine has no load connected to it (Figure 7.27). The turbine has the following dimensions: mass of rotor 500 kg; mean diameter of rotor 2·5 m; diameter of bearings 0·15 m; μ for the bearings 0·075. Calculate:

    a) The frictional torque.

    b) The force R exerted by the jet.

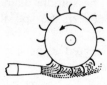

Fig. 7.27

18   In an electric motor, the two brushes are pressed against the commutator by springs which each exert a force of 2·5 N (Figure 7.28). Calculate μ between the brushes and commutator if the torque created is 37·5 Nmm.

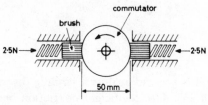

Fig. 7.28

19   For the brake mechanism shown (Figure 7.29), μ between the brake pad and drum is 0·65. Calculate the distance X in order to exert a braking torque of 12 Nm.

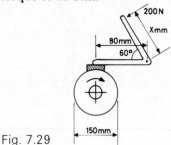

Fig. 7.29

20 The brake mechanism for a lathe is shown (Figure 7.30). Calculate:
  a) The force R_N exerted when a force of 125 N is applied on the
     pedal.
  b) The friction torque on the machine shaft, if $\mu = 0.55$.

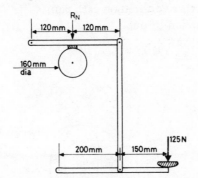

      Fig. 7.30

21 The tailstock of a lathe is to be used for drilling a hole in the end
    of a metal bar (Figure 7.31). If the axial force on the drill is 500 N
    and μ between the tailstock and lathe bed, and between lathe bed
    and clamp, is 0.15, calculate the clamping force to be exerted by the
    nut so that the tailstock will not slide.

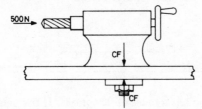

      Fig. 7.31

22 The ram of a shaping machine is pushed forward by connecting
    link AB (Figure 7.32). The resistance to cutting is 600 N, the mass
    of the ram is 45 kg and the coefficient of friction is 0.15 between the
    bearing surfaces. Find the force required in the link in the position
    shown.

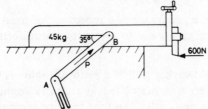

      Fig. 7.32

23 A cable operated brake is shown (Figure 7.33). The wheel diameter
    is 0·66 m and μ between the wheel rim and brake blocks is 0·7. If the
    pull in cable A is 180 N, determine:
    a) The force pressing each brake block against the wheel.
    b) The friction force exerted on each rim.
    c) The total friction torque.

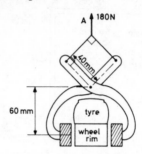

Fig. 7.33

24 Figure 7.34 shows a casting clamped to the table of a shaping
    machine. The casting has a mass of 14 kg, and is held down by
    forces of 1·9 kN at A and of 1·96 kN at B. The coefficient of friction
    between casting and table is 0·2. Find the maximum horizontal
    force P that may be exerted by the tool.

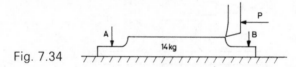

Fig. 7.34

25 A railway engine has a mass of 95 tonnes. If the coefficient of fric-
    tion between the wheels and rails is 0·1 calculate the maximum
    drawbar pull.

26 A lorry tows a compressor unit of mass 800 kg along a level road.
    The resistance to motion of the compressor is 55 N/100 kg. Calcu-
    late the pull required by the lorry.

27 Find the tractive resistance of a motor car of mass 650 kg which
    requires a force of 1·5 kN to move it at a uniform speed up an
    incline of 1 in 9.

28 The maximum drawbar pull of a railway engine is 95 kN. Deter-
    mine the greatest number of trucks that the engine may pull, if
    each truck has a mass of 11 t and the tractive resistance is 175 N/t.

29 A diesel locomotive draws a train of 55 trucks each of mass 12 t. The tractive resistance is 200 N/t. Calculate the drawbar pull on a level track. If the train runs down an incline of 1 in 200, calculate the pull now required.

30 A tractor of mass 1·5 t pulls a trailer loaded with straw bales (Figure 7.35). μ between the tractor wheels and the ground is 0·5. The trailer has a mass of 600 kg and the mass of each bale is 50 kg. Determine:
   a) The load on the driving wheels of the tractor.
   b) The maximum drawbar pull.
   c) The maximum number of bales carried if the tractive resistance of the trailer is 65 N/100 kg. Assume the load to be evenly distributed, i.e. no downward force on the tractor drawbar.

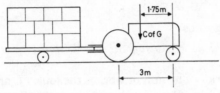

Fig. 7.35

31 The vice of a shaping machine exerts a clamping force of 6 kN on a block of steel (Figure 7.36). Calculate the maximum force P which may be exerted by the tool if μ between the contact surfaces is 0·2.

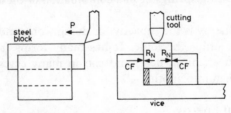

Fig. 7.36

8
Work Power and Energy

Work (W)   Work is done when a force moves its point of application through a distance, in the direction of the force; *i.e.* for a constant force:

$$\text{Work done} = \text{Force} \times \text{Distance moved}$$

or   *Work = Resistance overcome × Displacement*

i.e. $W = F \times S$

For a varying force:

Work done = Average Force × Distance

Unit of Work   The unit of work is the joule (J) and is the work done when a force of 1 newton moves through a distance of 1 metre.

i.e.   1 joule = 1 newton × 1 metre

1 J = 1 Nm

Work can be done against gravity, against friction, in stretching and compressing springs or in a combination of these ways.

Energy   Energy is the capacity for doing work, and is measured in the same unit—the joule. It has many forms (*e.g.* chemical, electrical, mechanical, heat) and can in many cases be changed from one form into another.

Mechanical Energy
1 *Potential* (*PE*) Due to the position of an object relative to a datum. Measured by the work that it can do in returning to the datum position.
2 *Kinetic* (*KE*) Due to the speed of an object. Can be measured by the work it will do against a resistance before being brought to rest.
3 *Strain* (*SE*) Due to deformation, as, for instance, the energy stored in a spring.

The total energy stored in a system is constant unless external work is supplied to, or taken from it.

Power (P)   Power is the rate of doing work. The unit of power is the watt (W) and is the rate of working when 1 joule is done in 1 second.

$$Power \text{ (watts)} = \frac{Total\ work\ done\ \text{(joules)}}{Total\ time\ taken\ \text{(seconds)}}$$

$$= Force \times \frac{Distance}{Time}$$

or $Power$ (watts) $= Force$ (newtons) $\times Speed$ (metres/sec.)

Work Done by a Torque   Let a force F at a radius r be rotated through N revolutions. Then for one revolution,

$$W = F(2\pi r)$$

and for N revolutions,

$$W = N \times 2\pi(Fr)$$

But,

$$T = Fr$$

$$W = 2\pi NT$$

If the force moves at n rev/s, the work done/second

$$= 2\pi nT$$

$$\therefore \quad P = 2\pi nT$$

Power Transmitted by Belts

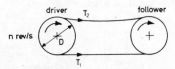

Fig. 8.1

Effective tension on the driver $= (T_1 - T_2)$
Hence work done/second $\quad = \pi Dn(T_1 - T_2)$ joules
Power transmitted $\qquad = \pi Dn(T_1 - T_2)$ watts

Belt Slip   If the belt does not slip then all the power will be transmitted to the follower. In most cases some slipping will occur at the pulley surfaces, the power loss being given as a percentage of the total.

Power Transmitted by Gears   Let the force on the teeth of
a gear rotating at n rev/s be F (newtons) at the pitch circle diameter.
Then the power transmitted,

   $P = \pi DnF$ (where D is the pitch circle diameter)

Worked Examples

1   Find the work done by a hydraulic car hoist in raising a car of mass
    900 kg to a height of 1·75 m (Figure 8.2).

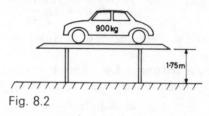

Fig. 8.2

$$Mass = 900 \text{ kg}, \quad \therefore \text{ Force of Gravity} = 900 \times 9\cdot81 \text{ N}$$
$$= 8810 \text{ N}$$
$$Work \text{ done} = \text{Force} \times \text{distance}$$
$$= 8810 \text{ N} \times 1\cdot75 \text{ m}$$
$$= 1541 \text{ J } or \text{ } 1\cdot54 \text{ kJ}$$

Hence work done in lifting the car is 1·54 kJ.

2   A lift of mass 500 kg moves upwards at a uniform rate of 0·4 m/s
    (Figure 8.3). The total frictional resistance of the guides amounts
    to 400 N. Calculate the power developed by the motor when the
    lift carries 2 people, each of mass 75 kg.

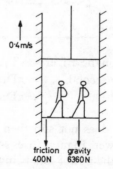

friction   gravity
400N   6360N

Fig. 8.3

$$\text{Friction force} = 400 \text{ N}$$
$$\text{Total mass} = (500 + 75 + 75) \text{ kg}$$
$$= 650 \text{ kg}$$
$$\therefore \quad \text{Gravity force} = 650 \times 9.81 \text{ N}$$
$$= 6\,360 \text{ N}$$
$$\text{Total force overcome} = \text{Friction force} + \text{gravity force}$$
$$i.e. \quad \text{Total force overcome} = 6\,760 \text{ N}$$
$$\text{Power} = \text{Force} \times \text{velocity}$$
$$= 6\,760 \times 0.4 \text{ W}$$
$$= 2\,704 \text{ W } or \text{ } 2.70 \text{ kW}$$

i.e. Power developed by the motor is 2·71 kW.

3   A winch drags a casting of mass 55 kg for a distance of 30 m across a workshop floor (Figure 8.4). The coefficient of friction between the casting and the floor is 0·63. Calculate:

a) The work done by the winch.

b) The average power developed, if the operation takes 25 seconds.

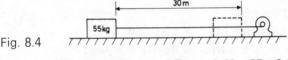

Fig. 8.4

$$\text{Frictional force} = \mu R_N = 0.63 \times 55 \times 9.8$$
$$= 340 \text{ N}$$
$$W = F \times S$$
$$= 340 \times 30 \text{ J}$$

i.e. Work done by the winch = 10 200 J

$$P = \frac{\text{Work done}}{\text{Time taken}}$$
$$= \frac{10\,200}{25} \text{ W}$$

i.e. Average power developed = 408 W.

4   A uniform turbine rotor of mass 150 kg, revolves in bearings of 80 mm diameter (Figure 8.5). If the rotor turns at 2 500 rev/min and μ between the bearing surfaces is 0·02, determine:

a) The load on each bearing.

b) The frictional force exerted on each bearing.

c) The work lost/minute in friction at each bearing.

d) The total power lost in overcoming friction.

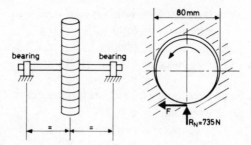

Fig. 8.5

a) Rotor is symmetrical ∴ half the mass supported on each bearing, i.e. 75 kg.

∴ *Load on each bearing* $= 75 \times 9 \cdot 81$ N

$$= 735 \ N$$

b) F = μRn

   $= 0 \cdot 02 \times 735$ N

i.e. Friction force $= 14 \cdot 7$ N

c) W = F $\times$ S

   $= 14 \cdot 7 \times \pi$dn

   $= 14 \cdot 7 \times \pi \times 80 \times 10^{-3} \times 2\,500$ Nm

i.e. Work lost at each bearing $= 9\,240$ J *or* $9 \cdot 24$ kJ

d) P $= \dfrac{\text{W}}{\text{t}}$

   $= \dfrac{9\,240 \times 2}{60}$ W (2 bearings)

i.e. Power absorbed in friction $= 308$ *watts*

5 An elevator carries bags from an unloading bay to a store (Figure 8.6). If a lorry load of 60 bags, each containing 50 kg is unloaded in 6 minutes, calculate the average power output of the motor. Assume all losses amount to 10% of the output.

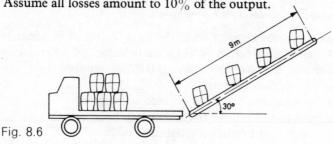

Fig. 8.6

Mass of each bag $= 50$ kg, thus the gravitational force (weight) of each bag $= 50 \times 9 \cdot 81 = 490$ N.

$$
\begin{aligned}
\text{Total W} &= \text{Total F} \times \text{S} \\
&= (490 \times 60) \times 9 \sin 30^\circ \text{ J} \\
&= 490 \times 60 \times 9 \times 0 \cdot 5 \text{ J}
\end{aligned}
$$

i.e. Total work $= 132 \cdot 3$ kJ

$$
P = \frac{W}{t} \times \frac{100}{90} \text{ (accounting for losses)}
$$

$$
P = \frac{132\,300}{6 \times 60} \times \frac{100}{90} \text{ watts}
$$

Average power output $= 408$ W

6   The power output of a car engine is 30 kW at 72 km/h on a level road. Find the total resistance to motion.

If the car has a mass of 800 kg, at what steady speed will it travel up an incline of 1 in 20, the tractive resistance and power output remaining constant?

$$
72 \text{ km/h} = 72 \times \frac{10}{36} \text{ m/s} = 20 \text{ m/s}
$$

$$
P = F \times v
$$

$$
\therefore \quad 30 \times 10^3 \text{ N} = F \times 20
$$

$$
F = \frac{30 \times 10^3}{20} \text{ N}
$$

Frictional resistance $= 1 \cdot 5 \times 10^3$ N

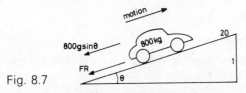

Fig. 8.7

$$
\text{Mass} = 800 \text{ kg}
$$

$$
\text{Weight} = 800 \times 9 \cdot 81 \text{ N}
$$

Total force to overcome $=$ Frictional resistance
                          $+$ gravity component (see Figure 8·7)

$$
= (1 \cdot 5 \times 10^3 + 800 \times 9 \cdot 81 \times \frac{1}{20}) \text{ N}
$$

$$
= 1\,500 \text{ N} + 392 \text{ N}
$$

$$
= 1\,892 \text{ N}
$$

$$P = F \times v$$

$$v = \frac{P}{F}$$

$$= \frac{30 \times 10^3}{1\,892} \text{ m/s}$$

$$= 15\cdot85 \text{ m/s}$$

$$= 15\cdot85 \times \frac{36}{10} \text{ km/h}$$

$$= 57 \text{ km/h}$$

Hence, the speed up the incline is 57 km/h.

7   Figure 8.8 shows a truck of mass 200 kg being hauled up an incline of 1 in 10 by a rope attached to a winch. The maximum power output of the winch is 12 kW, and the tractive resistance of the truck is 500 N/tonne. Calculate the maximum mass that can be hauled up the incline at a steady speed of 18 km/h.

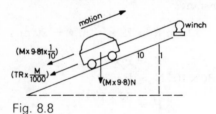

Fig. 8.8

$$\text{Speed} = 18 \text{ km/h} = 18 \times \frac{10}{36} \text{ m/s}$$

$$= 5 \text{ m/s}$$

Let Mkg be the total mass.

Force to overcome = Gravity component + frictional resistance

$$P = F \times v$$

$$12 \times 1\,000 = \left(\frac{M \times 9\cdot81}{10} + 500 \times \frac{M}{1\,000}\right) \times 5$$

(See sketch for total force)

$$M = \frac{12 \times 1\,000}{5(0\cdot98 + 0\cdot5)} \text{ kg}$$

$$= \frac{12\,000}{5 \times 1\cdot48} \text{ kg}$$

$$= 1\,620 \text{ kg}$$

But the mass of the truck is 200 kg.

∴ *Maximum mass is 1 420 kg.*

8   Water has to be pumped from a reservoir to a feeder tank 50 m
above the surface of the reservoir (Figure 8.9). If the tank is
10 m × 12 m × 4 m deep, calculate the work done in filling it.

How long would it take to fill if the power output of the pump is
20 kW?

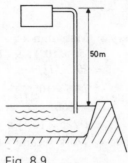

Fig. 8.9

$$\text{Volume of water} = 10 \times 12 \times 4 \text{ m}^3$$
$$= 480 \text{ m}^3$$
$$\text{Hence, mass of water} = 480 \times 10^3 \text{ kg}$$
$$W = F \times S$$
$$= (480 \times 10^3 \times 9 \cdot 81) \times 50 \text{ J}$$
$$= 235\,500 \text{ kJ}$$
$$\text{Power available} = 20 \text{ kW}$$
$$= 20 \text{ kJ/s}$$
$$\therefore \text{ Time taken} = \frac{235\,500}{20} \text{ seconds}$$
$$= \frac{235\,500}{20 \times 60} \text{ minutes}$$
$$= 196 \text{ minutes}$$

The time taken to fill the tank is thus 3 hours 16 minutes.

9   A Pelton Wheel is supplied with 50 m³ of water/min. from a reservoir whose surface is 150 m above the wheel (Figure 8.10). If the
overall efficiency of the system is 80%, calculate:
a) The total energy available/minute.
b) The power output from the wheel.
c) The torque at the shaft when the wheel rotates at 2 000 rev/min.

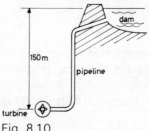

Fig. 8.10

a) *Total available energy* = Mass/min × g × height
   = $(50 \times 10^3) \times 9.81 \times 150$ J/min
   = *73.5 MJ/min.*

b)   Energy available = $\dfrac{73.5 \times 10^6}{60}$ J/s

   = 1.225×10^6 J/s

   Power output = Work/second × efficiency

   = $1.225 \times 10^6 \times \dfrac{80}{100}$ W

   i.e. Power output = 980 kW

c)   P = 2πnT

   ∴ $980 \times 10^3 = 2\pi \times \dfrac{2\,000}{60}$ T

   i.e. T = $\dfrac{980 \times 10^3 \times 60}{2\pi \times 2\,000}$

   i.e. Shaft torque = 4 680 Nm

10 A mass of 5·1 kg is hung from a spring of stiffness 0·5 N/mm (Figure 8.11). Calculate:
    a) The extension of the spring.
    b) The total mass which will extend the spring by a further 21·6 mm.
    c) The work done in extending the spring.

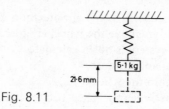

Fig. 8.11

a) Force on spring $= 5\cdot1 \times 9\cdot81$ N

$\qquad = 50$ N

$\qquad$ Stiffness $= \dfrac{\text{Force}}{\text{Extension}}$

$\qquad$ Extension $= \dfrac{\text{Force}}{\text{Stiffness}}$

$\qquad\qquad = \dfrac{50}{0\cdot5}$ mm

$\qquad\qquad = 100$ mm

b) $\qquad$ Force $=$ Extension $\times$ stiffness

$\qquad\qquad = 121\cdot6 \times 0\cdot5$ N

$\qquad\qquad = 60\cdot8$ N

$\therefore$ Mass $= \dfrac{60\cdot8}{9\cdot8}$ kg

The mass to stretch the spring by 121·6 mm = 6·2 kg.

c) $\qquad$ W $=$ Average force $\times$ distance

$\qquad\qquad = \dfrac{(0+60\cdot8)}{2} \times 121\cdot6$

$\qquad\qquad = 3\,700$ Nmm

Work done in stretching the spring = 3·7 J.

11 A machine of mass 150 kg is supported by four strong springs, one at each corner (Figure 8.12). If the machine has to compress the springs by 0·01 m when placed on them, calculate the stiffness of the springs.

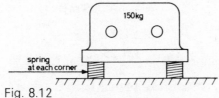

Fig. 8.12

Mass supported by each spring $= \dfrac{150 \text{ kg}}{4}$

$\therefore$ Force on each spring $= \dfrac{150}{4} \times 9\cdot81$ N

$\qquad$ Stiffness $= \dfrac{\text{Force}}{\text{Compression (change in length)}}$

$$= \frac{150 \times 9 \cdot 81}{4 \times 0 \cdot 01} \text{ N/m}$$

i.e. Spring stiffness = 36·7 kN/m

12 The sketches (Figure 8.13) show two plates held apart by two co-axial springs A and B. The stiffness of A is 7·5 N/mm and of B 1·75 N/mm. Calculate:
   a) The force F to maintain the assembly in the position shown in Figure 12(b).
   b) The work done against the springs.

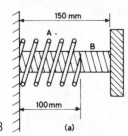 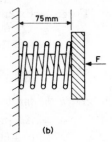

Fig. 8.13   (a)   (b)

a) Force to compress A = Stiffness × change in length
                        = 7·5 × 25 N
                        = 187·5 N

   Force to compress B = 1·75 × 75 N
                        = 131·25 N

   ∴ *Total force* = 318·75 N

b) Work to compress A = Average force × contraction

$$= \frac{187 \cdot 5}{2} \times 25 \text{ mJ}$$

$$= 2\,340 \text{ mJ}$$

   Work to compress B $= \dfrac{131 \cdot 25}{2} \times 75 \text{ mJ}$

$$= 4\,930 \text{ mJ}$$

Hence total work done against the springs is 7·27 J.

NOTE: Neither spring has any initial tension.

13 A lawnmower engine has a maximum power output of 800 W at 30 rev/s. The pulley diameter is 90 mm. Calculate:
   a) The tension in each side of the belt, at maximum power, if $T_1 = 2 \cdot 25 T_2$.

b) The power transmitted if the belt slip is 5%.

$$a) \qquad P = \pi Dn(T_1 - T_2)$$
$$= \pi Dn(2 \cdot 25T_2 - T_2)$$
$$P = \pi Dn \times 1 \cdot 25T_2$$

$$i.e. \quad T_2 = \frac{P}{1 \cdot 25\pi Dn}$$

$$= \frac{800}{1 \cdot 25 \times \pi \times 0 \cdot 09 \times 30} \, N$$

$$\therefore \quad T_2 = 75 \cdot 5 \, N$$
$$\text{and} \quad T_1 = 2 \cdot 25T_2$$
$$= 2 \cdot 25 \times 75 \cdot 5 \, N$$
$$= 170 \, N$$

Tensions in each side of the belt are 170 N and 75·5 N.

$$b) \qquad \text{Power output} = 800 \text{ watts}$$
$$\text{Belt slip} = 5\%$$
$$i.e. \text{ Power lost} = 5\%$$
$$\therefore \text{ Power transmitted} = 95\% \text{ of } 800 \text{ watts}$$
$$= \frac{95}{100} \times 800 \, W$$
$$= 760 \, W$$

Examples

1   Calculate the work done in raising 8 barrels each of mass 51 kg on to the back of a lorry 1·25 m above the ground.

2   A casting of mass 500 kg is moved a distance of 15 m along a workshop floor. Find the work done if μ between the surfaces is 0·7.

3   Calculate the work done by a force of 25 N when it moves its point of application through a distance of 6 m.

4   A force of 56 kN moves through a distance of 0·6 m. Calculate the work done.

5   Determine the work done by a tug in moving a cargo vessel 2·5 km upstream to a berth, when the average tension in the hawser is 85 kN.

6   Find:
    a) The work done in raising a mass of 51 kg to a height of 2·5 m.
    b) The power required if this takes 5 seconds.

7   In moving a distance of 5 m, a force does 3·75 kJ of work. Calculate:
   a) The value of the force.
   b) The power output if the time taken is 20 seconds.

8   Find the power output of a lift motor which can raise a cage of mass 750 kg to a height of 20 m in 30 s.

9   A fork lift truck raises a barrel of oil onto the back of a lorry. The height of the lorry floor is 1·55 m and the barrel has a mass of 185 kg. Calculate the work done by the fork lift.

10   A builder's hoist is used to lift 25 concrete blocks from ground level 8·5 m to the second floor of a building. The blocks each have a weight of 250 N. Calculate:
   a) The work done.
   b) The power output if the hoist takes 29·5 s to travel from ground level to the second floor.

11   The power rating of a motor is 500 watts. Calculate the work which can be done by the motor in 2 minutes.

12   A hydraulic ram has a stroke of 0·3 m and exerts a force of 4 500 N (Figure 8.14). Find:
   a) The work done/stroke.
   b) The power developed if the ram makes 15 strokes/minute.

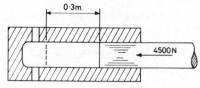

Fig. 8.14

13   A torque of 12 Nm moves a shaft through 15 revolutions. Find the work done by the torque.

14   A vee pulley is driven by a belt which exerts an effective tension of 80 N. The diameter of the pulley is 120 mm. Calculate:
   a) The torque on the pulley.
   b) The power transmitted at 24 rev/s.

15   The power output of a portable electric drill is 300 W. Assuming the efficiency of the drill to be 75%, calculate the torque exerted on a 6 mm drill which rotates at 1 100 rev/min.

16 Determine the power of a pump used to raise 95 litres of water to a height of 16 m in 30 s. What is the potential energy of the water in the final position? (1 litre of water has a mass of 1 kg.)

17 An electric motor driving a trolley has a power rating of 700 W. Calculate:
   a) The distance that the trolley can be moved in 1 second against a resistance of 350 N.
   b) The time taken to move the trolley 15 m.

18 An overhead crane lifts a machine of mass 250 kg to a height of 5 m in 6·5 s (Figure 8.15). Calculate the power output of the crane motor if the efficiency is 75%.

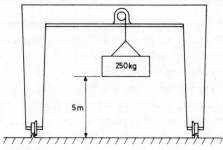

250kg

5m

Fig. 8.15

19 The stiffness of a spring is 4 N/mm. Determine:
   a) The force required to stretch the spring by 10 mm.
   b) The average force exerted during the process.
   c) The work done in stretching the spring.
   d) The energy stored in the stretched spring.

20 A uniform beam for a bridge has to be lifted 75 mm into position by means of a hydraulic jack (Figure 8.16). The mass of the beam is 840 kg. Calculate:
   a) The work done by the jack.
   b) The minimum diameter of the piston if the pressure in the fluid is not to exceed 15 N/mm^2.

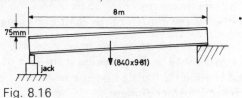

8m

75mm

jack

(840×9·81)

Fig. 8.16

21   When the foundations of a power station are being dug water soaks into the pit at a rate of 2 500 litres/minute. If the water has to be lifted 6 m out of the pit, calculate the power output of the pump, in order to keep the pit dry.

22   A hydro-electric station is supplied with water from a dam at a total head of 150 m. The power output of the station is to be 50 MW. The losses due to friction in the pipeline and energy remaining in the water leaving the turbine amount to 15% of the available energy. Calculate:

a) The mass of water used/second.

b) The volume of water used/minute.

c) The power lost.

23   A flywheel of mass 24 kg is supported in bearings as shown in Figure 8.17. If the value of the coefficient of friction between the shaft and the bearings is 0·05, calculate:

a) The frictional force at each bearing.

b) The frictional torque if the shaft is 25 mm in diameter

c) The frictional work done in bringing the flywheel to rest if it makes 20 revolutions in coming to rest.

d) The power absorbed in friction if this takes 15 seconds.

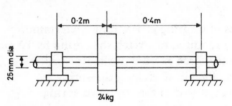

Fig. 8.17

24   A shaft of mass 229·5 kg and diameter 0·1 m rotates in bearings at each end. μ between the shaft and bearings is 0·05. Calculate:

a) The frictional torque.

b) The work done in bringing the shaft to rest when, at a certain speed, the kinetic energy of the shaft is 3 kJ.

c) The number of revolutions of the shaft in coming to rest.

25   A tension spring, under an initial tension of 160 N has to be stretched a further 0·14 m. If the spring stiffness is 6 kN/m, calculate the final tension. Calculate also the work done in stretching the spring by this further amount.

26 A brake pedal has a return spring as shown in Figure 8.18. The pedal moves 60 mm to apply the brake, and the spring stiffness is 6·5 kN/m. Determine:

a) The tension in the spring in this position if the initial tension is 25 N.

b) The work done against the spring.

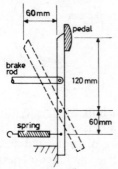

Fig. 8.18

27 The tup of a pile driver has a mass of 300 kg, and is raised 3·5 m for each stroke (Figure 8.19). Calculate:

a) Its potential energy in the raised position.

b) Its energy just before it hits the pile, if the frictional resistance amounts to 150 N.

c) The average power output of the driving motor when the pile driver makes 12 strokes/minute.

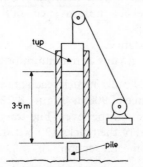

Fig. 8.19

28 Figure 8.20 shows a car brake drum. If the coefficient of friction between the brake drum and the brake lining is 0·6, and the hydraulic system exerts a force of 150 N, determine:

a) The frictional force created.

b) The frictional torque.

The wheel rotates 14 times between the brake being applied and the car coming to rest. Find:

c) The work done against friction.

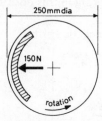

Fig. 8.20

29 A shunting engine is pushing two trucks into a siding (Figure 8.21). The trucks each have a mass of 12 t and the tractive resistance is 150 N/t. The trucks are disconnected and roll for 200 m before coming to rest. Calculate the kinetic energy of the trucks at the instant of disconnection.

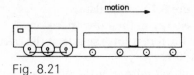

Fig. 8.21

30 41 kW is the maximum power output of a car, whose mass is 825 kg. If it can travel at a uniform speed of 81 km/h up an incline of 1 in 6, determine:

a) The tractive resistance.

b) The work done if it runs at maximum power for 3 minutes.

31 An engine of mass 90 t pulls 3 trucks each of mass 15 t up an incline of 1 in 60 for a distance of 900 m (Figure 8.22). It then pulls them for a distance of 2 km on the level. The tractive resistance of the engine is 300 N/t and of the trucks 120 N/t. Determine:

a) The work done on each section.

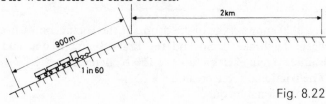

Fig. 8.22

b) The power output of the engine when it travels up the incline at a speed of 54 km/h.

c) The speed on the level with the same power output.

32 Calculate the power transmitted by a pulley 0·15 m diameter rotating at 600 rev/min, when the difference in belt tensions between the slack and tight sides is 150 N.

33 A pulley 0·1 m in diameter is to be used to transmit 650 W at 15 rev/s. Find the required tensions in each side of the belt, given that $T_1 = 2\frac{1}{2}T_2$.

34 The vee pulley driving a circular saw rotates at 1 400 rev/min and has a diameter of 0·2 m. Find the power of the motor required if the belt tensions are: $T_1 = 258$ and $T_2 = 105$ N.

35 The motor and pump of a washing machine are connected by a belt drive, the motor pulley having a diameter of 60 mm and the pump pulley a diameter of 90 mm. The tensions in the two sides of the belt are 128 N and 71 N, and the motor output is 250 W. Determine:

a) The speed of the motor.

b) The speed of the pump assuming no slip.

c) The speed of the pump assuming 5% slip.

36 A screwcutting lathe is driven from a 1·5 kW motor by three vee belts. The motor speed is 24 rev/s and the pulley diameter is 125 mm. Determine the tensions in each belt if $T_1 = 2·25\,T_2$.

9
Work Diagrams

Since work is the product of the two quantities force and displacement, and area is the product of the two quantities length and breadth, then work can be represented by an area.

If we draw the graph of force against displacement for a *constant* force, as shown in Figure 9.1 then the area under the graph to the 3 m mark is $2 \times 3 = 6$ square units. The work done is $2 \times 3 = 6$ J.

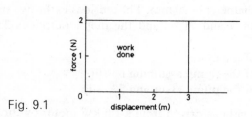

Fig. 9.1

If the force *varies linearly*, then the graphs will be as shown in Figure 9.2(a), (b) and (c).

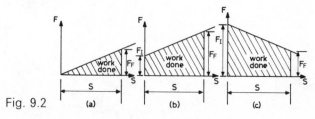

Fig. 9.2   (a)   (b)   (c)

When the force is increasing uniformly as in (a) and (b), the graph slopes upward to the right. When there is no initial force then the graph starts at the origin, but if there is an initial force (*e.g.* initial tension on a spring) then the graph is as shown in (b) with the initial force F_1. With a decreasing force the graph slopes downward to the right, as in sketch (c).

In each of these cases, the work performed = average force × displacement.

$$i.e. \quad W = \frac{(F_I + F_F)}{2} \times S$$

The area of a trapezium = half the sum of the parallel sides × the perpendicular distance between them, which is of the same form as the work equation and thus once again the area under the graph can represent the work done. The area can be found either by using the trapezoidal formula, or simply by dividing the diagram into rectangles and triangles.

The work diagrams for actual practical situations may be built up from a number of parts as illustrated by some of the following worked examples. There are, however, many cases where the force does not vary uniformly and more complex methods, beyond the scope of this book, must be used.

Worked Examples

1 A mine shaft is 250 m deep. A hoist, of mass 2·6 t rises empty from the bottom to a gallery at 210 m, below the surface where it is loaded with 3 full hutches, each of mass 0·6 t and an empty hutch of mass 0·25 t. The empty hutch is unloaded from the hoist at a second gallery 60 m above the first one. Draw a work diagram to show the total work done in raising the hoist from the foot of the shaft to the surface.

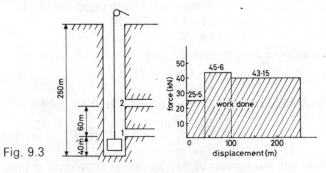

Fig. 9.3

Initial force = 2·6 × 1 000 × 9·81 N
= 25·5 × 10³ N

Force added at ① = [(3 × 0·6) + 0·25]1 000 × 9·81 N

$$= 2 \cdot 05 \times 9 \cdot 81 \times 10^3 \text{ N}$$
$$= 20 \cdot 1 \times 10^3 \text{ N}$$

$\therefore$ Total force at ① $= 45 \cdot 6 \times 10^3 \text{ N}$

Force subtracted at ② $= 0 \cdot 25 \times 1\,000 \times 9 \cdot 81 \text{ N}$
$$= 2 \cdot 45 \times 10^3 \text{ N}$$

$\therefore$ Force after ② $= 43 \cdot 15 \times 10^3 \text{ N}$

Work done $=$ Total shaded area
$$= [(25 \cdot 5 \times 10^3 \times 40) + (45 \cdot 6 \times 10^3 \times 60)$$
$$+ (43 \cdot 15 \times 10^3 \times 150)] \text{ J}$$
$$= (1\,020 + 2\,740 + 6\,460) \text{kJ}$$

i.e. Work done = 10 220 kJ or 10·22 MJ

2   A chain of mass $1 \cdot 02$ kg/m hangs from a drum as shown in the sketch (Figure 9.4). Draw a work diagram, and find:
   a) The work done in winding the chain onto the drum.
   b) The work done in raising the first 5 m.

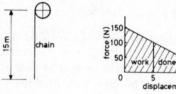

Fig. 9.4

Initial force $=$ Length (m)
$$\times \text{mass/unit length (kg/m)} \times 9 \cdot 81 \text{ N}$$
$$= 15 \times 1 \cdot 02 \times 9 \cdot 81 \text{ N}$$
$$= 150 \text{ N}$$

a) Total work done $=$ Shaded area
$$= \tfrac{1}{2} \times 150 \times 15 \text{ J}$$
$$= 1 \cdot 125 \text{ kJ}$$

b) Force required after 5 m are raised $= 100 \text{ N}$
$$\therefore \quad \text{W} = \frac{(150 + 100)}{2} \times 5 \text{ J} = 625 \text{ J}$$

3   A passenger lift has a mass of 510 kg and is supported by two cables, each having a mass of $0 \cdot 765$ kg/m. The lift shaft is 20 m deep. Draw a work diagram showing the work done in raising the cage from the bottom to the top of the shaft.

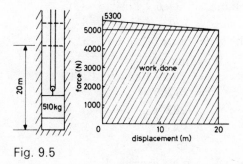

Fig. 9.5

$$\text{Initial force exerted on lift} = 510 \times 9 \cdot 81 \text{ N}$$
$$= 5\,000 \text{ N}$$
$$\text{Initial force exerted on rope} = 2 \times 20 \times 0 \cdot 765 \times 9 \cdot 81 \text{ N}$$
$$= 300 \text{ N}$$
$$\therefore \text{ Total initial force} = 5\,300 \text{ N}$$
$$\text{Final force} = 5\,000 \text{ N}$$
$$W = \text{Area shaded}$$
$$= \frac{(5\,300 + 5\,000)}{2} \times 20 \text{ J}$$
$$= 103\,000 \text{ J} \quad i.e. \quad 103 \text{ kJ}$$
$$or \quad W = \text{Area of rectangle} + \text{area of triangle}$$
$$= (5\,000 \times 20 + \tfrac{1}{2} \times 300 \times 20) \text{ J}$$
$$= 100\,000 \text{ J} + 3\,000 \text{ J}$$
$$= 103 \text{ kJ}$$

4   The level of a canal lock, when full, is 4 m above the level of the canal (Figure 9.6). The lock is 6 m wide and 30 m long. The lock has to be filled in 30 minutes. Calculate:
  a) The total mass of water required.
  b) The average power of the pump: (i) when the pipe discharges at the top of the lock; (ii) when the pipe discharges half way up.
  c) Draw a work diagram of the filling process in each case.

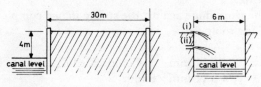

Fig. 9.6

a) Total volume $= 30 \times 6 \times 4 \ m^3$

$\therefore$ *Total mass* $= 720 \times 10^3 \ kg$

b) (i) All the water must be raised 4 m.

$$\therefore \quad W = mg \times 4$$
$$= (720 \times 10^3) \times 9 \cdot 81 \times 4 \ J$$
$$= 28 \ 200 \ kJ$$

$$\therefore \quad P = \frac{282}{3 \times 6} \ kW$$
$$= 15 \cdot 7 \ kW$$

(ii) $W = mg \times$ average distance

$$= 720 \times 10^3 \times 9 \cdot 81 \times \frac{(4+2)}{2} \ J$$
$$= 720 \times 10^3 \times 9 \cdot 81 \times 3 \ J = 21 \ 200 \ kJ$$
$$P = \frac{21 \ 200}{30 \times 60} \ kW = 11 \cdot 75 \ kW$$

c)

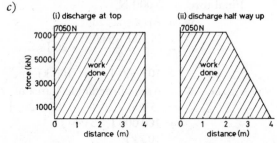

Fig. 9.7

5   A locating device for the toolpost of a lathe consists of a cam and a spring loaded plunger as shown (Figure 9.8). The spring strength is 4·5 N/mm and there is no initial compression.

a) Calculate the force exerted by the spring in its fully compressed position.

b) By drawing a work diagram find the work done in compressing the spring.

c) Find the work done against the spring in one complete revolution of the toolpost.

a) Stiffness $= 4 \cdot 5$ N/mm   Compressed distance $= 8$ mm

$$F = 4 \cdot 5 \text{ N/mm} \times 8 \text{ mm}$$
$$F = 36 \cdot 0 \text{ N}$$

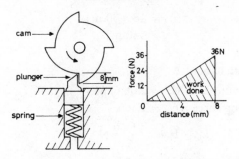

Fig. 9.8

b) $\qquad$ W = Shaded triangle

$$= \frac{(0+36)}{2} \times 8 \times 10^{-3} \text{ J}$$

$$\therefore \quad W = 144 \text{ mJ}$$

c) There are four "teeth" on the cam and thus the spring will be compressed four times in one complete revolution.

The work done in one revolution is thus $4 \times 144 \text{ mJ} = 576 \text{ mJ}$.

6 An accelerator adjustment rod has a tension return spring (Figure 9.9). The initial tension is 6 N and the spring stiffness is 0·8 N/mm.

Draw a work diagram to show the work done when the spring is stretched 3 cm. Calculate this work and also the effort F required at B to maintain the position.

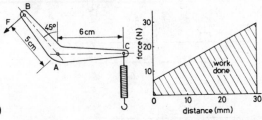

Fig. 9.9

$\qquad$ Initial tension = 6 N $\qquad$ Increase in tension = $0·8 \times 30$ N
$$= 24 \text{ N}$$

$$\therefore \quad \text{Final tension} = 6 \text{ N} + 24 \text{ N} = 30 \text{ N}$$

$$W = \frac{(F_I + F_F)}{2} S$$

$$= \frac{(6+30)}{2} \times 30 \text{ Nmm}$$

$$= 18 \times 30 \text{ mJ}$$
$$= 540 \text{ mJ}$$

To find F:

Take moments about A

For equilibrium Σ C.M. $= \Sigma$ A.C.M.

(Spring force $\times 6$) $= F \times 5$

$$i.e. \quad F = \frac{30 \times 6}{5} \text{ N}$$

$$i.e. \quad F = 36 \text{ N}$$

$\therefore$ *A force of 36 N is required at B to stretch the spring by the given amount.*

7   In a foundry, a ladle which runs on rails as shown (Figure 9.10) is used to pour molten metal into the moulds. The mass of ladle and contents when full is 510 kg, and it is used to fill three moulds of 100 kg, 250 kg and 75 kg mass respectively. Draw a work diagram and determine the work done in filling the moulds if the frictional resistance is 0·6 N/kg and the ladle moves 30 m from the furnace to the first mould.

How much work is required to return the ladle to the furnace?

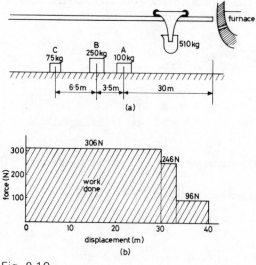

Fig. 9.10

Initial force to be overcome $= 510 \times 0.6$ N $= 306$ N
Reduction in force at A $= 100 \times 0.6$ N $= 60$ N
Reduction in force at B $= 250 \times 0.6$ N $= 150$ N
Reduction in force at C $= 75 \times 0.6$ N $= 45$ N

W $=$ Shaded area $= (30 \times 306) + (246 \times 3.5) + (96 \times 6.5)$
$= 9\,180 + 860 + 624 = 10.66$ kJ

Work done in returning the ladle $= 51 \times 40 = 2.04$ kJ

Examples

1   A compression spring has a free length of 80 mm and a stiffness of
6 N/mm. Calculate the force required to compress it to 65 mm in
in length. Draw a work diagram and calculate:
a) The work done in compressing the spring by this amount.
b) The work done in the first 10 mm of compression.

2   The suspension system of a trailer consists of a coiled spring connected by a short pivoted beam (Figure 9.11). The mass supported
by each wheel is 500 kg and the spring stiffness is 0.04 kN/mm.
Draw a work diagram for the compressed spring, and determine
the work done.

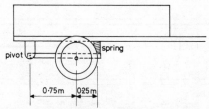

Fig. 9.11

3   A pit cage of mass 2 000 kg is carried by a wire rope of mass 4.8 kg/m.
Find, by means of a work diagram, the work done in raising the
cage from the bottom of a shaft 90 m deep.

4   Draw a diagram, and calculate the work done in winding the first
20 m of cable onto a drum, if 30 m of cable are hanging freely and
it has a mass of 3.06 kg/m.

5   Figure 9.12 shows the cross section of a well containing water.
Draw a diagram to show the work done in emptying the well, and
find the total work done. Find also the work done in pumping the
first 6.0 m^3.

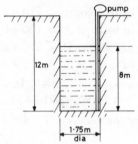

Fig. 9.12

6 A lift of mass 350 kg carried 3 people of average mass 75 kg from
   the ground floor to the second floor, 8 m above. It then proceeds
   empty a further 12 m to the fifth floor. By means of a work diagram
   find the total work done in reaching the fifth floor. Neglect the
   effects of friction and weight of the rope.

7 An overhead conveyor supplies components to an assembly line
   (Figure 9.13). The conveyor which runs on overhead rails has a
   mass of 30 kg, the frictional resistance being 0·5 N/kg. It carries
   16 components, each of mass 2·5 kg. Draw a work diagram and
   find the work done in supplying one component to each of the
   first five stations, the stations being 2 m apart.

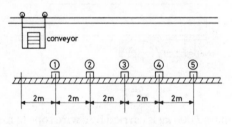

Fig. 9.13

8 A manure spreader (Figure 9.14) has a mass of 350 kg. When
   loaded it carries a mass of 450 kg. The resistance to motion is
   1 500 N/tonne and the spreader travels 600 m in distributing its
   entire load. Draw a work diagram and determine the total work
   done in moving the spreader and its load. On a second run of
   600 m, after travelling 450 m, the spreader is reloaded with a further
   250 kg, but still spreads at the same rate. Draw a diagram and find
   the work done on this run.

mass of manure
450kg

mass of spreader
350kg

Fig. 9.14

9 A pump is used to lift 25·5 m³ of water from a reservoir to a tank 30 m above the water surface. Determine by means of a work diagram the work done during the process. (Assume the level of the reservoir to remain unaltered.)

10 A skip of concrete is lifted 15 m by a tower crane, from ground level to the point where it is to be poured (Figure 9.15). Draw a work diagram and find the work done during the operation, if the mass of the skip is 153 kg. Neglect the mass of the rope.

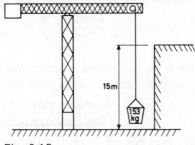

15m

153 kg

Fig. 9.15

11 A tension spring is used in the pick up arm of a record player as part of the switching mechanism. It is 3 cm long when under an initial tension of 0·2 N. It is stretched to a length of 5 cm at the point when it switches the player off. The spring stiffness is 0·005 N/mm. Determine from a work diagram the work done in stretching the spring.

10
Machines

A machine is a device whereby an energy input can be modified and delivered as an output in a form more suitable for a particular purpose.

In most—but not all—cases, a large load will be moved a small distance by a small effort moving through a large distance.

The following three quantities are defined in machine theory.

1 Mechanical Advantage is the ratio of load to effort.

$$i.e. \ \ M.A. \ = \ \frac{Load}{Effort} \ = \ \frac{W}{P}$$

2 Velocity Ratio is the ratio of the distance moved by the effort to the distance moved by the load in the same time.

$$i.e. \ \ VR = \frac{Distance \ moved \ by \ effort}{Distance \ moved \ by \ load} = \frac{Dp}{Dw}$$

The velocity ratio depends solely on the construction, and is a constant for any particular machine.

3 Efficiency is the ratio of work output to work input.

$$i.e. \ \ Efficiency = \frac{Work \ output}{Work \ input}$$

$$or \ \ \ \eta = \frac{WO}{WI}$$

where η (the Greek letter eta) is used to denote efficiency. Often the efficiency is quoted as a percentage. This gives:

$$\eta \ per \ cent \ = \ \frac{WO}{WI} \times 100$$

$$\eta = \frac{WO}{WI} = \frac{W \times Dw}{P \times Dp}$$

$$= \frac{W}{P \times \left(\dfrac{Dp}{Dw}\right)}$$

$$i.e. \quad \eta = \frac{W}{P \times VR}$$

$$or \quad \eta = \frac{MA}{VR}$$

It should be noted that although the last two equations are useful for solving problems, they are *not* a definition of efficiency. They are derived from it.

The Ideal Machine   The ideal machine is one where the work output is equal to the work input. This is never realised in practice since some work must always be done to overcome friction and the weight of the machine (*i.e.* WO = WI − losses). The idea, however, of an ideal machine is sometimes used for calculating the losses.

If the work output = work input, then $\eta = 1$ (in all other cases $\eta < 1$).

Thus $\eta = \dfrac{MA}{VR} = 1$ and MA = VR for an ideal machine.

Calculation of Velocity Ratios   This should be done, as far as possible, from first principles. The method is to let the load move unit distance and then calculate the corresponding distance moved by the effort. Figure 10.1 gives a reference list for some of the common machines.

Technical terms used with various machine elements
1 Pulleys
a) *Block :* a block is a number of pulleys assembled in a supporting framework.
b) *Block and Tackle :* the complete lifting machine, comprising two blocks and a rope.
c) *Snatch block :* a snatch block is a single pulley attached to the load. It may either be used alone or as part of a more complex machine. In each case it doubles the VR.

MACHINE	SKETCH	VELOCITY RATIO	η
lever	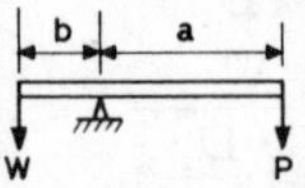	length of effort arm / length of load arm $\dfrac{a}{b}$	high
simple wheel and axle	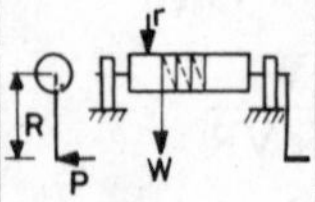	radius of handle / radius of axle $\dfrac{R}{r}$ or $\dfrac{D}{d}$	high
block and tackle (various numbers of pulleys)	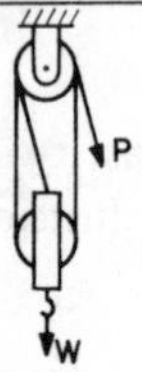	number of ropes supporting the lower block or number of pulleys in use N	high
wedge	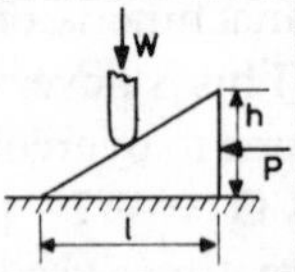	horizontal distance / vertical distance $\dfrac{l}{h}$	medium
inclined plane	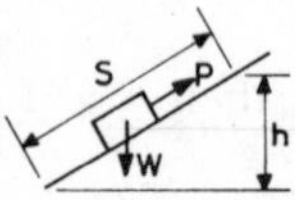	length of slope / vertical height $\dfrac{s}{h}$	medium
differential wheel and axle	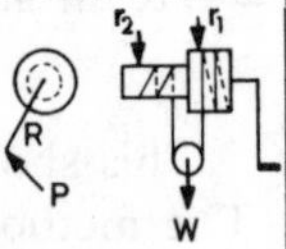	radius of handle / half difference of radii of axles $\dfrac{R}{\frac{1}{2}(r_1 - r_2)}$ or $\dfrac{D}{\frac{1}{2}(d_1 - d_2)}$	high
Weston differential pulley	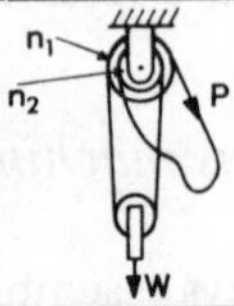	flats on large pulley / half difference of flats on large and small $\dfrac{n_1}{\frac{1}{2}(n_1 - n_2)}$	low
screw jack	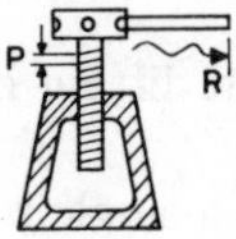	circumference of effort handle / pitch of screw $\dfrac{2\pi R}{P}$	low

Fig. 10.1 (continued on next page)

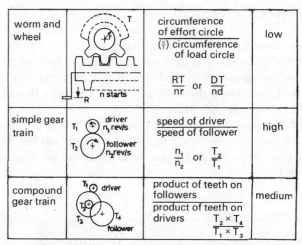

worm and wheel		circumference of effort circle $\overline{(\frac{n}{T})}$ circumference of load circle $\dfrac{RT}{nr}$ or $\dfrac{DT}{nd}$	low
simple gear train	driver n_1 rev/s T_1 / follower n_2 rev/s T_2	speed of driver speed of follower $\dfrac{n_1}{n_2}$ or $\dfrac{T_2}{T_1}$	high
compound gear train	T_1 driver T_3 T_2 T_4 follower	product of teeth on followers product of teeth on drivers $\dfrac{T_2 \times T_4}{T_1 \times T_3}$	medium

Fig. 10.1 (continued)

2 Gears (Figure 10.2)

a) *Pitch Circle:* the circle at which two gears may be taken as touching. The diameter of this circle is called the pitch circle diameter (P.C.D.).

b) *Circular Pitch:* the distance between one tooth and the corresponding point on the next tooth, measured round the pitch circle.

c) *Idler:* any gear which comes between the driver and the follower. An idler does not alter the velocity ratio, but changes the direction of rotation.

d) *Simple Gear Train* (Figure 10.2(b)): when each gear in a series meshes directly with the next one, this is known as a simple gear train.

e) *Compound Gear Train* (Figure 10.2(c)): when two of the gears in a train are fixed rigidly (keyed) to the same shaft, then the arrangement is known as a compound gear train.

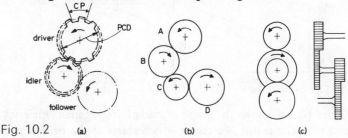

Fig. 10.2   (a)   (b)   (c)

3 Screws (Figure 10.3)

a) *Pitch:* the pitch of a screw thread is the distance from a point on one thread to the corresponding point on the next thread.

b) *Number of Starts:* the most usual form is to have one thread cut on the surface of the rod, but, if it is required that the nut move rapidly along the thread, then two or more threads may be cut alongside each other.

c) *Lead:* the lead is the distance that the nut will move along the thread in one revolution. For a single start thread the lead will be equal to the pitch. For an 'n' start thread the lead will be n × pitch.

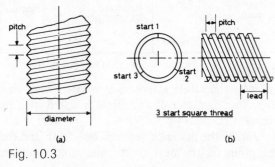

(a)   (b)

Fig. 10.3

Load-Effort and Load-Efficiency Graphs If an experiment is carried out on a machine and a graph of load plotted against the effort required, it is found that the resulting graph is a straight line (Figure 10.4(a)). It should be noted however that because of experimental error all of the points will not lie exactly on the line. This means that values may not be taken from the table of results but should be taken from the graph for any further calculations.

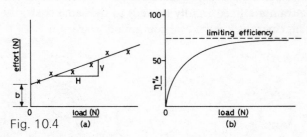

Fig. 10.4   (a)   (b)

Figure 10.4(b) shows the graph of load plotted against efficiency. In this case the result is a curve which starts at the origin, rises steeply at first and then gradually levels off. The maximum value which would be reached with a very large load is known as the limiting efficiency.

The Law of the Machine Since the load-effort graph is a straight line, it has the mathematical form $y = mx + c$. The equation connecting the load and the effort is called the law of the machine and is given the symbolic form:

$$P = aW + b$$

where 'a' is the gradient of the line ($\dfrac{V}{H}$ in Figure 10.4)

'b' is the effort required for no load.

Limiting Efficiency

Since $\eta = \dfrac{W}{P \times VR}$ and $P = aW + b$

then $\eta = \dfrac{W}{(aW + b)VR} = \dfrac{W}{aW \times VR + b \times VR}$

$= \dfrac{1}{aVR + \dfrac{bVR}{W}}$ (on dividing top and bottom by W)

As the load becomes very large, $\dfrac{b \times VR}{W}$ becomes very small

i.e. $\dfrac{b \times VR}{W} \to 0$ and $\eta \to \dfrac{1}{aVR}$ hence the limiting value of $\eta = \dfrac{1}{aVR}$

Overhauling In a lifting machine if, when the effort is removed, the machine runs back, it is said to overhaul. For any machine
$WI = WO + \text{Friction losses}$
i.e. Friction losses $= WI - WO$.
If the machine is to overhaul then the load must overcome the friction.

i.e. $\qquad WO > \text{Friction losses}$

$\qquad WO > WI - WO$

$\qquad \dfrac{WO}{WI} > 1 - \dfrac{WO}{WI}$, dividing throughout by WI.

i.e. $\qquad \eta > 1 - \eta$

and this can only be the case if $\eta > 0.5$

∴ if η is greater than 50% the machine will overhaul

if η is less than 50% the machine will be self sustaining (will not run back).

Worked Examples

1 It is found when using a certain machine that a load of mass 75 kg requires a force of 49 N to raise it.

Determine:

  a) The mechanical advantage.

  b) The work output when the load is raised 2·5 m.

  a) *Mechanical advantage* $= \dfrac{\text{Load}}{\text{Effort}}$

  $= \dfrac{75 \times 9.81 \, \text{N}}{49 \, \text{N}} = 15.0$

  b) *Work output* $=$ Force $\times$ distance moved

  $= 75 \times 9.81 \times 2.5 \, \text{J}$

  $= 1.84 \, \text{kJ}$

2 Find the distance that a load may be raised by a machine of velocity ratio 12, when the effort moves a distance of 4 m.

$$\text{VR} = \frac{\text{Distance moved by effort}}{\text{,, \quad ,, \quad ,, \ load}}$$

$$12 = \frac{4 \, \text{m}}{\text{Dw}}$$

$$\text{Dw} = \frac{4}{12} \, \text{m} = 0.33 \, \text{m}$$

3 The VR of a machine is 9. Calculate the efficiency if a load of mass 66 kg can be lifted by an effort of 120 N.

$$\eta = \frac{\text{WO}}{\text{WI}}$$

$$\therefore \eta = \frac{\text{W}}{\text{P} \times \text{VR}} = \frac{66 \times 9.81}{120 \times 9} = 0.6$$

4 In a lifting machine, a load of 1 000 N can be overcome by an effort of 150 N. The load is raised 0·5 m when the effort moves through a distance of 6 m. Calculate:

a) The velocity ratio.
b) The mechanical advantage.
c) The work input.
d) The work output.
e) Efficiency.
f) Work lost.

a) $Velocity\ ratio = \dfrac{Dp}{Dw} = \dfrac{6}{0\cdot5} = 12$

b) $Mechanical\ advantage = \dfrac{W}{P} = \dfrac{1\,000}{150} = 6\cdot675$

c) $\begin{aligned}Work\ input &= P \times Dp \\ &= 150 \times 6\ J \\ &= 900\ J\end{aligned}$

d) $\begin{aligned}Work\ output &= W \times Dw \\ &= 1\,000 \times 0\cdot5\ J \\ &= 500\ J\end{aligned}$

e) $\eta = \dfrac{WO}{WI} = \dfrac{500}{900} = 0\cdot555$

f) $\begin{aligned}Work\ lost &= WI - WO \\ &= (900 - 500)\ J = 400\ J\end{aligned}$

5 A block and tackle has 3 pulleys in the top block and 2 in the bottom block. Calculate:
a) The load which can be lifted by an effort of 120 N if the efficiency at this load is 80%.
b) The work output if the load is raised a distance of 0·75 m.

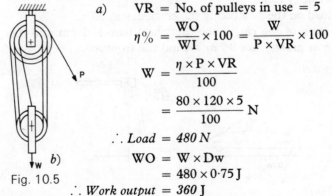

a) $VR = \text{No. of pulleys in use} = 5$

$\eta\% = \dfrac{WO}{WI} \times 100 = \dfrac{W}{P \times VR} \times 100$

$\begin{aligned}W &= \dfrac{\eta \times P \times VR}{100} \\ &= \dfrac{80 \times 120 \times 5}{100}\ N\end{aligned}$

$\therefore Load = 480\ N$

b) $\begin{aligned}WO &= W \times Dw \\ &= 480 \times 0\cdot75\ J\end{aligned}$

$\therefore Work\ output = 360\ J$

Fig. 10.5

6 The drum of a wheel and axle (windlass) has a diameter of 150 mm (Figure 10.6). The effort required to lift a load of 770 N is 110 N. Calculate:

a) The efficiency percent, if the length of the handle is 0·6 m.

b) The work done by the effort in moving the load 4·5 m.

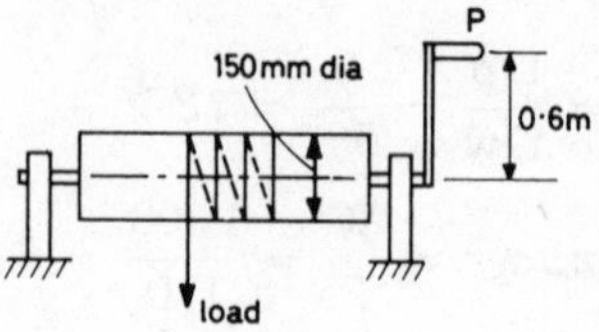

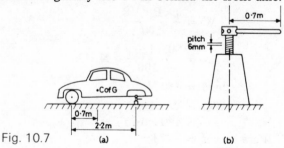

Fig. 10.6

a)
$$VR = \frac{D}{d} = \frac{2 \times 0.6 \text{ m}}{0.15} = 8$$

$$\eta = \frac{WO}{WI} = \frac{W}{P \times VR} = \frac{770}{110 \times 8}$$

$$\eta = 87.5\%$$

b)
$$D_P = VR \times D_W$$
$$= 8 \times 4.5 \text{ m}$$
$$\therefore \text{Distance moved by effort} = 36 \text{ m}$$
$$WI = P \times D_P$$
$$= 110 \times 36 \text{ J}$$
$$= 3.96 \text{ kJ}$$

7 A screw jack is used to lift the rear axle of a car (Figure 10.7). The pitch of the screw is 6 mm and the effort of 125 N is applied at the end of a handle 0·7 m long. Determine:

a) The velocity ratio.

b) The load on the jack taking the efficiency as 30%.

c) The mass of the car, if the car wheel base is 2·2 m and the centre of gravity lies 0·7 m behind the front axle.

Fig. 10.7   (a)   (b)

a)
$$VR = \frac{\text{Distance moved by effort}}{\text{,,   ,,   ,, load}} = \frac{2\pi R}{P} \text{ for screw jack}$$

$$VR = \frac{2\pi \times 0.7}{6 \times 10^{-3}}$$

i.e.   VR = 734

b)
$$\eta\% = \frac{WO}{WI} = \frac{W}{P \times VR} \times 100$$

$$W = \frac{\eta \times P \times VR}{100}$$

$$= \frac{30 \times 125 \times 734}{100} \text{ N}$$

$$\therefore W = 27.5 \text{ kN}$$

c) Taking moments about the front axle
$$(mg \times 0.7) = 2.2 \times 27\,500$$

$$m = \frac{2.2 \times 27\,500}{0.7 \times 9.81} = 882$$

i.e. The mass of the car is 882 kg.

8 The results of a test on a machine of VR 12 are shown in the table below.

LOAD (N)	200	400	600	800	1 000	1 200	1 400
EFFORT (N)	50	75	100	120	140	165	190

Draw the load-effort graph and determine:
a) The law of the machine.
b) The effort required for a load of 900 N.
c) The efficiency at this load.

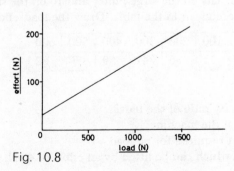

Fig. 10.8

a) Choose two points on the graph.
   When W = 500 N, P = 85 N
   and W = 1 500 N, P = 200 N
   Substitute these values in the equation $P = aW + b$

$$85 = a500 + b \quad (1)$$
$$200 = a1\,500 + b \quad (2)$$

   Subtract (1) from (2)
$$115 = a1\,000 + 0$$

$$a = \frac{115}{1\,000} = 0.115$$

   Substitute for 'a' in equation (1)
$$85 = 0.115 \times 500 + b$$
$$b = 85 - 57.5$$
   i.e. $b = 27.5$

∴ *The law of the machine is* $P = 0.115\,W + 27.5$.

NOTE: An alternative method is given in Question 9.

b)   If W = 900,
$$P = 0.115 \times 900 + 27.5$$
$$= 103.5 + 27.5$$
   ∴ P = 131 N

c) $$\eta = \frac{WO}{WI} = \frac{W}{P \times VR}$$

   i.e. $$\eta = \frac{900}{131 \times 12}$$

   i.e. $\eta = 0.572$ for this load

9 A test was carried out on a Weston differential pulley block which has 11 flats on the large pulley and 10 on the small pulley. The results are given in the table. Draw the load-effort graph.

LOAD(N)	100	200	300	400	500	600
EFFORT(N)	25	36	47	59	70	82

Find:
a) The velocity ratio of the block.
b) The law of the machine.
c) The effort required for no load.
d) The load which can be lifted by an effort of 40 N.

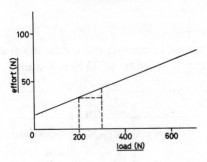

Fig. 10.9

a) VR of Weston's Pulley $= \dfrac{n_1}{\frac{1}{2}(n_1 - n_2)}$

$$= \dfrac{2 \times 11}{(11 - 10)}$$

$$= \dfrac{22}{1} = 22$$

b) From the graph, 'b' (the intercept on the effort axis) = 13. To find 'a', choose two points on the graph. It helps in calculation if the loads are chosen to give an even number since this will be the denominator of the gradient. Take W = 200 and W = 300, P = 36 and P = 47.

$$a = \dfrac{V}{H} = \dfrac{47 - 36}{300 - 200} = \dfrac{11}{100} = 0{\cdot}11$$

∴ *The law of the machine is P = 0·11 W + 13*

c) The effort required for no load is 'b' (*i.e.* where the 'zero load' cuts the graph).

∴ *The effort required for no load is 13 N*

d) When P = 40 N, 40 = 0·11 W + 13

$$0{\cdot}11\,W = 40 - 13$$

$$W = \dfrac{27}{0{\cdot}11}$$

∴ *The load = 245 N*

10 A rack and pinion is used to move the saddle along the bed of the lathe (Figure 10.10). The mass of the saddle is 15 kg and μ between the bearing surfaces is 0·1. The details of the mechanism are: rack: pitch of teeth 6 mm; pinion: 16 teeth; radius of handle circle: 75 mm.

Calculate:

a) The velocity ratio.

b) The work output in moving the saddle 0·5 m along the slides.

c) The efficiency if an effort of 33 N is required.

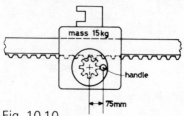

Fig. 10.10

a) $$\text{VR} = \frac{\text{Distance moved by effort}}{\text{,,} \quad \text{,,} \quad \text{,,} \quad \text{load}}$$

Let the effort handle make 1 complete revolution. Then the effort moves $2\pi R = 2\pi \times 75$ mm. In the same time the load (frictional resistance) moves 16 teeth $= 16 \times 6$ mm

$$\therefore \text{VR} = \frac{2\pi \times 75}{16 \times 6} = 4\cdot9$$

b) Load = Frictional resistance = μR_N
$$= 0\cdot1 \times (15 \times 9\cdot81)$$

i.e.   Load (W) = 147 N

Work output = $W \times Dw = 147 \times 0\cdot5$

Work output = *73·5 J*

c)
$$\eta = \frac{\text{WO}}{\text{WI}} = \frac{\text{W}}{\text{P} \times \text{VR}}$$

$$\eta = \frac{147}{33 \times 4\cdot9}$$

i.e. $\eta = 0\cdot91$

11 A single purchase winch is used to lift a concrete slab of mass 95 kg (Figure 10.11). The winch drum has a diameter of 160 mm, the large gear has 180 T and the pinion has 30 T. An effort of 55 N is required at the end of the handle, which has a length of 0·3 m. Calculate:

a) The velocity ratio of the winch.

b) The efficiency % at this load.

c) The tension in each of the slings A and B.

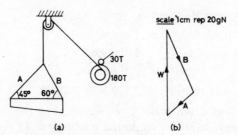

Fig. 10.11   (a)   (b)

a) Let the load drum make 1 revolution.

∴ Effort handle makes $\frac{180}{30}$ revolutions = 6 revs.

$$VR = \frac{\text{Distance moved by effort}}{\text{,, \quad ,, \quad ,, \quad load}}$$

$$= \frac{(6 \times 2\pi \times 0.3)\,m}{(1 \times \pi \times 160 \times 10^{-3})\,m}$$

$$= 22.5$$

b) $\eta\% = \frac{WO}{WI} \times 100$

$$= \frac{W}{P \times VR} \times 100$$

$$= \frac{95 \times 9.81 \times 100}{55 \times 22.5}$$

∴ $\eta = 75.3\%$

c) From the triangle of forces (Figure 10.11(b)), B = 3.45 cm and A = 2.45 cm.

∴ *Tension in B = 675 N and A = 480 N*

2 A double purchase crab winch is used to hold a beam in position as shown in Figure 10.12. The mass of the beam is 450 kg and the details of the winch are given in the sketch. Calculate:

a) The tension T in the rope.
b) The velocity ratio of the winch.
c) The effort required if the efficiency is 0.7.

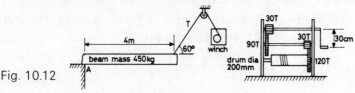

Fig. 10.12

a) Taking moments about A

$$(450 \times 9\cdot81) \times 2 \ = \ T \sin 60° \times 4$$

$$T \ = \ \frac{450 \times 9\cdot81 \times 2}{0\cdot866 \times 4} \ N$$

$\therefore$ *Tension in the rope* $= 2\,550\,N$ *or* $2\cdot55\,kN$

b) Let the load drum make 1 revolution then the handle makes $\frac{120}{30} \times \frac{90}{30}$ revolutions (compound gearing) = 12 revs.

$$VR \ = \ \frac{\text{Distance moved by effort}}{\text{,,} \qquad \text{,,} \quad \text{,,} \quad \text{load}}$$

$$= \ \frac{12 \times 2\pi R}{1 \times \pi D} \ = \ \frac{12 \times 2\pi \times 0\cdot3\,(\text{m})}{1 \times \pi \times 0\cdot2\,(\text{m})}$$

$$\therefore VR \ = \ 36$$

c)
$$\eta \ = \ \frac{W}{P \times VR} \qquad P \ = \ \frac{W}{\eta \times VR}$$

$$= \ \frac{2\,550}{0\cdot7 \times 36} \ N$$

$$= \ 101\,N$$

Effort required to hold the beam in position is 101 N.

13 A motor shaft, rotating at 12 rev/s has an 80 T gear A fixed to it. This meshes with a gear on another shaft B which is required to rotate at 16 rev/s. Calculate:

a) The number of teeth on the second gear.

b) The P.C.D. of each of the gears, the circular pitch being 6 mm.

a)
$$\frac{N_A}{N_B} \ = \ \frac{T_B}{T_A}$$

$$T_B \ = \ \frac{N_A}{N_B} \times T_A$$

$$= \ \frac{12}{16} \times 80 \ = \ 60$$

Number of teeth on wheel B is 60.

b) Circular pitch $\times$ number of teeth $= \pi \times$ pitch circle diameter

$$\text{C.P.} \times T \ = \ \pi \times \text{P.C.D.}$$

$$\text{P.C.D. for } A \ = \ \frac{\text{C.P.} \times T}{\pi}$$

$$= \frac{6 \times 80}{\pi} \text{ mm} = 153 \text{ mm}$$

$$\text{and } P.C.D. \text{ for } B = \frac{6 \times 60}{\pi} \text{ mm} = 114 \cdot 5 \text{ mm}$$

14 The dimensions of a worm and wheel used to raise a sluice gate are shown in Figure 10.13. The load on the machine is 5·75 kN and an effort of 160 N is required to overcome this. Determine:

a) The mechanical advantage.

b) The velocity ratio.

c) The efficiency.

d) The output power if the gate is raised 1·5 m in 50 seconds.

e) The power of an electric motor which could replace the effort at A.

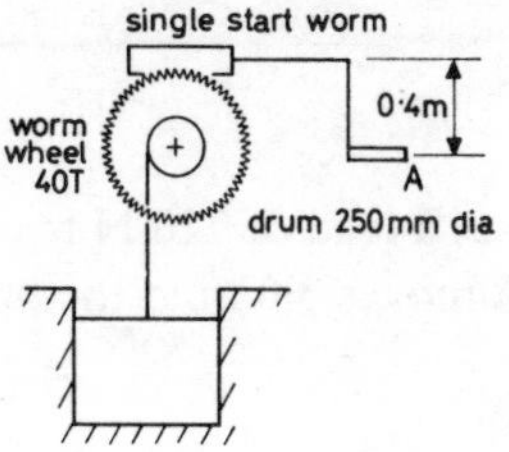

Fig. 10.13

a) $$\text{MA} = \frac{W}{P}$$

$$= \frac{5\ 750}{160}$$

$$= 36$$

b) $$\text{VR} = \frac{\text{RT}}{\text{rn}}$$

$$= \frac{0 \cdot 4 \times 40}{125 \times 10^{-3} \times 1}$$

$$= 128$$

c) $$\eta = \frac{\text{WO}}{\text{WI}} = \frac{\text{MA}}{\text{VR}} = \frac{36}{128} = 0 \cdot 281$$

d) $$\text{Power} = \text{Work done/s} = \frac{W \times D_W}{t} = \frac{5 \cdot 75 \times 10^3 \times 1 \cdot 5}{50} \text{ W}$$

$$= 173 \text{ Watts}$$

e) $$\text{Power input} = \frac{\text{Power output}}{\eta} = \frac{173}{0 \cdot 281} \text{ W} = 616 \text{ watts}$$

∴ *The motor would need to develop 616 watts.*

Examples

1   In a certain machine, the effort moves through a distance of 7·5 m when the load is moved 1·5 m. Calculate the velocity ratio.

2   The effort required to lift a load of 625 N is 125 N. Determine the mechanical advantage and the efficiency if the velocity ratio is 6.

3   A particular machine can lift a load of 2·5 kN through a distance of 0·25 m when the effort of 130 N moves through a distance of 5·5 m. Find:
   a) Mechanical advantage.
   b) Velocity ratio.
   c) Work output.
   d) Work input.
   e) The efficiency.

4   A lifting machine raises a load of 720 N to a height of 0·4 m. The effort, a force of 90 N, moves 3·6 m in the process. Calculate:
   a) Work output.
   b) Work input.
   c) Work lost.
   d) η

5   A load of mass 38·25 kg is raised 3·5 m by an effort of 50 N. The efficiency is 60%. Determine:
   a) The velocity ratio.
   b) The distance moved by the effort.
   c) Work output.

6   A block and tackle is used to lift a casting of mass 51 kg. Each block has 2 pulleys, and an effort of 150 N is required. Calculate:
   a) Velocity ratio.
   b) The mechanical advantage.
   c) The efficiency percent.
   d) The work input when the casting is raised 1·5 m.

7   A simple wheel and axle is used for closing the curtains at the front of a stage. The handle has a radius of 160 mm and the drum has a diameter of 150 mm. The resistance offered by the curtain is 55 N and an effort of 30 N is needed to overcome this. The curtain is pulled back 3·25 m. Calculate:

a) VR.
b) WO.
c) WI.
d) η.

8   An 8° wedge is used to raise the corner of a machine. The mass of
    the machine is 200 kg, the position of its centre of gravity being
    shown in the sketch (Figure 10.14). Find:

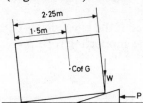

Fig. 10.14

a) The load on the wedge.
b) The velocity ratio of the wedge.
c) The effort P if the efficiency is 25%.

9   A box, of mass 25·5 kg is slid up an inclined ramp on to a lorry
    (Figure 10.15). The force P required is 147 N. Calculate:
    a) The work input in raising the box up the incline.
    b) The work output.
    c) The efficiency of the inclined plane.
    d) The coefficient of friction between the plane and the box.

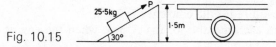

Fig. 10.15

10 A concrete foundation block is moved by a 3-2 block and tackle
    (Figure 10.16). It requires an effort of 110 N to move the block
    which has a mass of 102 kg. The coefficient of friction between the
    ground and the block is 0·4. Calculate:
    a) The velocity ratio.
    b) The efficiency.
    c) The work done by the effort in moving the block 2·75 m.

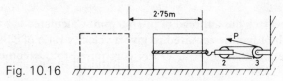

Fig. 10.16

11 A pulley block is to be used to lift a load of 550 N with an effort of 120 N. Taking the maximum efficiency as 87·2%, determine:
a) The velocity ratio.
b) The number of pulleys in each block.

12 Figure 10.17 shows a wall crane with a snatch block arrangement for lifting barrels, of mass 51 kg, from a lorry. The beam is uniform and has a mass of 76·5 kg. Determine:
a) The velocity ratio.
b) The effort required if the efficiency is 0·9.
c) The tension in the supporting rope A.

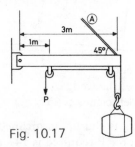

Fig. 10.17

13 The engine of a car is lifted out by means of a Weston pulley block (Figure 10.18). The block which has 9 flats on the large pulley and 8 on the small pulley is supported by an A pole. The mass of the engine is 81·5 kg. Calculate:
a) The velocity ratio.
b) The effort required if the efficiency is 28%.
c) The force in each member of the A pole when the engine is suspended from it.

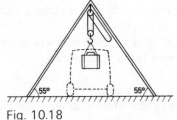

Fig. 10.18

14 A simple screw jack has a thread of pitch 6 mm. Calculate:
a) The length of handle required to give a velocity ratio of 314.
b) The efficiency if an effort of 175 N is required to overcome a load of 19·3 kN.

15 The turnbuckle shown (Figure 10.19) is used for tightening the mast stays of a yacht. It consists of a central body and two screws, each of pitch 1·5 mm. One screw has a right hand thread and the other has a left hand thread. The ends of the turnbuckle are prevented from rotating and it is tightened by a tommy bar of effective radius 143 mm, rotating about the centre line.
Calculate:
a) The velocity ratio of the device.
b) The tension in the stay if the effort exerted is 80 N and the efficiency at this load is 25%.

NOTE: For 1 revolution of the body, *each* screw will move through 1 pitch.

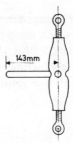

Fig. 10.19

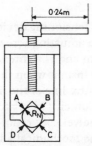

Fig. 10.20

16 A pipe vice is shown (Figure 10.20). The pitch of the screw is 4 mm and the handle radius is 0·24 m. Calculate:
a) The velocity ratio.
b) The vertical force exerted when an effort of 175 N is applied to the handle, and the efficiency is 36%.
c) The force exerted on the pipe at A, B, C, and D *i.e.* the normal reactions.
d) The maximum torque which can be applied to the pipe (diameter 25 mm) without it turning, if μ between vice and pipe is 0·25.

17 A simple lever press used for punching holes in sheet metal 1·5 mm thick (Figure 10.21) requires an effort of 160 N. The metal offers a resistance of 2·1 kN. Calculate:
a) The velocity ratio.
b) The distance that the effort will move when the punch moves 1·5 mm.
c) The work input.

d) The work output i.e. the work required to punch the hole.
e) The efficiency percent.

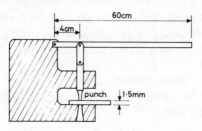

Fig. 10.21

18 A belt tensioning device for a motor consists of a hinged mounting
and an adjusting screw S (Figure 10.22). The screw has a pitch of
1·5 mm and is tightened by a spanner of effective radius 15·9 cm.
At the instant when the load on the system is 2 kN (acting through
the centre line of the motor shaft), an effort of 10 N is required on
the spanner. Calculate:
a) The velocity ratio.
b) The mechanical advantage.
c) The efficiency.

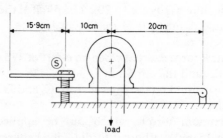

Fig. 10.22

19 A worm and wheel hoist has a two start worm and a 50 tooth worm
wheel. The load drum is 0·175 m in diameter and the effort handle
has a radius of 0·2 m. A load of 1·6 kN requires an effort of 90 N.
Calculate:
a) V.R.
b) M.A.
c) η.

0 On a test of a block and tackle, the following results were obtained.

LOAD (N)	100	200	300	400	500	600
EFFORT (N)	29	41	57	70	84	99

Draw the load-effort graph and determine:
a) The law of the machine.
b) The effort required for a load of 550 N.

21 The following results were obtained from a laboratory test on a machine of V.R. 7.

LOAD (N)	50	100	150	200	250
EFFORT (N)	16	25	34	40	48

Draw the load-effort graph and determine:
a) The law of the machine.
b) The effort required for a load of 225 N.
c) The corresponding efficiency.

2 The following two sets of values were taken from the load-effort graph of a screw jack of velocity ratio 50.

$$\text{when } W = 500 \text{ N,} \qquad P = 60 \text{ N}$$
$$\text{when } W = 1\,000 \text{ N,} \qquad P = 96 \text{ N}$$

Calculate:
a) The law of the machine.
b) The mechanical advantage for a load of 1 250 N.
c) The efficiency at this load.
d) The limiting efficiency.

3 A test on a 3-2 block and tackle supplied the results shown in the table.

LOAD (N)	100	350	450	550	800
EFFORT (N)	40	115	140	165	235

Draw the load-effort graph and determine:
a) The law of the machine.
b) The limiting efficiency.
c) The efficiency for a load of 600 N.

24 A rack and pinion is used as the feeding arrangement of a vertical drilling machine (Figure 10.23(a)). A force of 165 N applied to the handle creates a force of 1·4 kN at the drill point. Find:
a) The velocity ratio.
b) The efficiency percent.
c) The work output if the drillpoint moves 10 mm against the resistance.
d) The increase in effort required if the force is applied as shown in Figure 10.23(b).

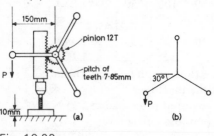

Fig. 10.23

25 A shaft with a 30T gear wheel rotates at 400 rev/min. This is geared to a parallel shaft which has to rotate at 160 rev/min. Calculate the number of teeth on the gear of the second shaft, and the P.C.D. if the circular pitch is 5 mm.

26 Calculate the six velocity ratios available with the following three gear wheels: 25T, 30T, 45T.

27 The speed of a lathe headstock pulley (Figure 10.24) is 260 rev/min. Calculate the speed of the chuck when the back gear cluster A, B, C, D, in Figure 10.24 is in mesh. The gears have the following numbers of teeth: A–30T; B–56T; C–21T; D–65T.

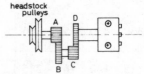

Fig. 10.24

28   The sawtable in a sawmill is traversed by a rack and compound
      gear as shown in Figure 10.25. The pitch of the teeth on the rack is
      20 mm. Calculate:
      a) The velocity ratio of the mechanism.
      b) The load which can be overcome if the efficiency is 0·58 and
         an effort of 120 N is applied to the handle.

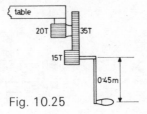

Fig. 10.25

29   A single purchase winch is used to drag a boat 6 m up a slipway.
      The tension in the rope is 1 600 N and an effort of 213 N is required
      to move the boat. The details of the winch are as follows:
      pinion–30 teeth; gear wheel–150 teeth; rope drum–30 cm dia;
      effort handle–25 cm radius.
      Calculate:
      a) The velocity ratio.
      b) The work output.
      c) The work input.
      d) The work lost.

30   The arrangement of a double purchase geared winch is shown in
      Figure 10.26. If the efficiency is 72%, calculate:
      a) The velocity ratio.
      b) The mass of the load which can be raised by an effort of 142 N
         applied at the handle.

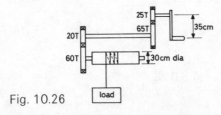

Fig. 10.26

31   A winch is used to hoist a 110 kg transformer into a position on
      an 'H' pole as shown in Figure 10.27(a). The details of the winch
      are shown in (b). Find:

a) The tension in the rope.
b) The reaction at the pulley A.
c) The velocity ratio of the winch.
d) The effort required if the $\eta = 0.65$.
e) What would happen if the effort were removed?

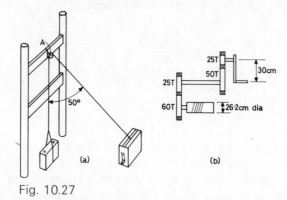

(a)   (b)

Fig. 10.27

32   The winch shown (Figure 10.28) is driven by an electric motor.
     The load, of mass 51 kg, is raised 3·5 m in 7 seconds. Calculate:
     a) The velocity ratio of the gearing.
     b) The speed of the drum.
     c) The speed of the motor.
     d) The power output.
     e) The power developed by the motor if the winch $\eta = 85\%$.

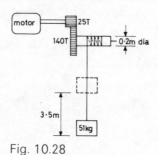

Fig. 10.28

11
Engines

Work Done in Cylinder The pressure (P) in an engine cylinder (Figure 11.1) is not constant throughout the stroke and thus the force on the piston is not constant. The average force on the piston = the average pressure × area of the piston

$$i.e. \quad F_A = P_M \times A$$

The average pressure is given the name *Mean Effective Pressure* (P_M).

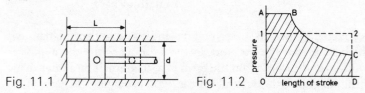

Fig. 11.1   Fig. 11.2

$$\text{Work done/stroke} = \text{Average force} \times \text{length of stroke}$$
$$= P_M \times A \times L$$
$$\text{and Work done/second} = P_M \times A \times L \times N$$
$$= P_M LAN \text{ (joules)}$$

where P_M = mean effective pressure (N/m^2); A = area of piston (m^2); L = length of stroke (m); N = number of working strokes/second.

Thus, Power developed = $P_M LAN$ (watts)
Note also: $P = P_M(LA) N$
$$= P_M VN \text{ where V is the 'swept'}$$
$$\text{volume of the cylinder.}$$

Number of Working Strokes The number of working strokes per revolution varies for different types of engine.

If N is the number of working strokes per second and n is the engine speed in revolutions per second:
1 For a single acting steam engine N = n.

2 For a double acting steam engine $N = 2n$ (*i.e.* 2 every rev.).
3 For a four stroke internal combustion engine $N = n/2$ (*i.e.* 1 every 2 revs.).
4 For a two stroke internal combustion engine $N = n$.

Mean Effective Pressure The mean effective pressure is that pressure which, if it acted throughout the working stroke would have the same effect as the actual pressure. Figure 11.2 shows a simplified graph of pressure-length of stroke for a steam engine, in which A,B,C, is the actual pressure during the stroke and 1,2, is the mean effective pressure. ABCDO represents, to a scale, the work done during the stroke.

$$\therefore \text{ Area ABCDO } = \text{ Area 12DO}$$
$$\text{Area ABCDO } = \text{ OD} \times \text{mean effective pressure}$$

i.e.   Mean effective pressure $= \dfrac{\text{Area under curve}}{\text{Length of base}}$

Indicator Diagrams An engine indicator is an instrument which measures the pressure in the cylinder while the engine is working, and records it in the form of a diagram. The actual shape of the diagram will depend on the type of engine (Figure 11.3). Because of the inertia of the moving parts, mechanical indicators are not used for speeds above 1 000 rev/min.

The area of the diagram is the shaded area (a), and the mean height of the diagram = Area/length.

The actual height of the diagram depends on the particular spring used in the indicator, the 'spring number' being the pressure/unit height of the diagram.

i.e.   Mean effective pressure = Mean height × spring number.

$$P_M = \frac{\text{Area}}{\text{Length}} \times S(N/m^2)$$

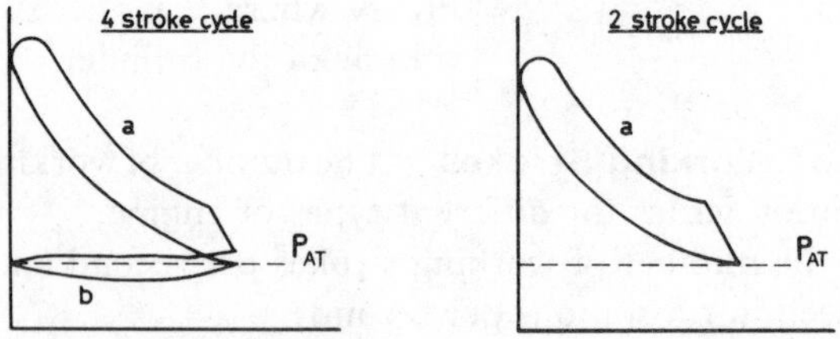

Fig. 11.3 Area (b) the pumping loop is small in comparison to (a) and is usually neglected

The area of the diagram may be found by the mid-ordinate rule or by using a planimeter. The power given by the indicator diagram is called the *indicated power* (IP).

$$i.e. \quad IP = P_M LAN$$

This is for one cylinder only and must be multiplied by the number of cylinders for multi-cylinder engines.

Brake Power (BP) The brake power is that available to do useful work, and is less than the indicated power because of friction losses in the engine. It may be measured by a brake test, hence the name brake power.

The Prony Brake This brake, normally used at low speeds, is shown in Figure 11.4. The two halves have a friction lining and are clamped lightly onto the shaft. When the shaft is rotating the friction torque is exactly balanced by $W \times L$.

$$\text{Thus BP} = 2\pi nT$$
$$= 2\pi n(WL) \text{ watts}$$

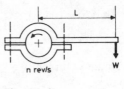

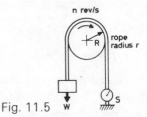

Fig. 11.4   Fig. 11.5

The Rope Brake In this type (Figure 11.5) the load is supported by a rope which passes over the flywheel, the other end being attached to a spring balance. The effective force on the rim of the flywheel is therefore $(W - S)$ where S is the spring balance reading. The torque on the flywheel is $(W - S)(R + r)$ and this balances the friction torque.

$$\therefore \quad BP = 2\pi nT$$
$$i.e. \quad BP = 2\pi n(W - S)(R + r)$$

If P_L is the power lost in the engine, then:

$$BP = IP - P_L$$

and the mechanical efficiency of the engine $= \dfrac{\text{Power output}}{\text{Power input}}$

$$i.e. \quad \eta \text{ mech} = \frac{\text{BP}}{\text{IP}}$$

Worked Examples

1   An indicator card had an area of $3\,600\,\text{mm}^2$ and a length of $90\,\text{mm}$. If the indicator spring number was $12{\cdot}5 \times 10^3\,\text{N/m}^2/\text{mm}$, calculate the mean effective pressure.

$$P_M = \frac{A}{L} \times S$$

$$= \frac{3\,600}{90} \times 12{\cdot}5 \times 10^3\,\text{N/m}^2$$

i.e.   *Mean effective pressure* $= 0{\cdot}5 \times 10^6\,N/m^2$

2   In a certain engine, the piston makes a stroke of $60\,\text{mm}$, and the bore of the cylinder is $75\,\text{mm}$.

    Calculate the work done during one stroke if the mean effective pressure is $0{\cdot}75 \times 10^6\,\text{N/m}^2$.

$$P_M = 0{\cdot}75 \times 10^6\,\text{N/m}^2,\ L = 60 \times 10^{-3}\,\text{m},\ A = \frac{\pi}{4} \times 75^2 \times 10^{-6}\,\text{m}^2$$

$$= 4\,420 \times 10^{-6}\,\text{m}^2$$

$$\dot{W} = P_M LA = 0{\cdot}75 \times 10^6 \times 60 \times 10^{-3} \times 4\,420 \times 10^{-6}\,\text{J}$$

Work done $= 199\,joules$

3   Find the indicated power output of a single cylinder two stroke engine, given the following information:
piston area $3\,500\,\text{mm}^2$; stroke $100\,\text{mm}$; mean effective pressure $0{\cdot}86\,\text{N/mm}^2$; engine speed $3\,000\,\text{rev/min}$;

$$P_M = 0{\cdot}86 \times 10^6\,\text{N/m}^2,\ L = 100 \times 10^{-3}\,\text{m},\ A = 3\,500 \times 10^{-6}\,\text{m}^2$$

$$N = \frac{3\,000}{60} = 50 \text{ working strokes/second}$$

$$IP = P_M LAN$$
$$= 0{\cdot}86 \times 10^6 \times 100 \times 10^{-3} \times 3\,500 \times 10^{-6} \times 50\,\text{W}$$
$$= 0{\cdot}86 \times 3\,500 \times 5\,\text{watts}$$
$$= 15{\cdot}05\,\text{kW}$$

4   The power output of a 4 cylinder 4 stroke engine has to be $30\,\text{kW}$ at $5\,000\,\text{rev/min}$. The stroke is $60\,\text{mm}$ and the piston area $4\,400\,\text{mm}^2$.

Determine the mean effective pressure required.

Power/cylinder $= \dfrac{30}{4}$ kW $= 7\cdot5$ kW, N $= \dfrac{5\,000}{60} \times \dfrac{1}{2} = 41\cdot6$

$$IP = P_M\,LAN$$

$$Thus: P_M = \dfrac{IP}{LAN}$$

$$= \dfrac{7\cdot5 \times 100}{60 \times 10^{-3} \times 4\,400 \times 10^{-6} \times 41\cdot6}\ N/m^2$$

Required mean effective pressure $= 0\cdot682 \times 10^6\ N/m^2$

5   The following information was taken from a test on an engine: engine speed 420 rev/min; length of arm $1\cdot5$ m; mass of the arm $9\cdot13$ kg.

Determine the brake power of the engine.

$$BP = 2\pi n\,T$$
$$= 2\pi n W\,L$$
$$= 2\pi \times \dfrac{420}{60} \times 9\cdot13 \times 9\cdot81 \times 1\cdot5\ \text{watts}$$
$$= 5\,900\ \text{watts}$$

$\therefore$ *Brake power* $= 5\cdot9\,kW$

6   A rope brake was used in a test on a gas engine.

Determine the brake power given the following details: engine speed 360 rev/min; W $= 400$ N; spring balance reading 50 N; diameter of brake $0\cdot98$ m; rope diameter 20 mm.

n $= \frac{360}{60} = 6$ rev/s, W $= 400$ N, S $= 50$ N, D $= 0\cdot98$ m, d $= 0\cdot02$ m

$$BP = 2\pi n(W-S)(R+r)$$
$$= \pi n(W-S)(D+d)$$
$$= \pi \times 6(400-50)(0\cdot98+0\cdot02)\ W$$
$$= \pi \times 60 \times 350 \times 1\ W$$

i.e.   BP $= 6\,600$ watts

$\therefore$ *Brake power of the engine is* $6\cdot6\,kW$

7   A single cylinder internal combustion (i.c.) engine working on the four stroke cycle has a stroke of 300 mm and a cylinder bore of 230 mm. The flywheel has a diameter of $1\cdot48$ m and the brake rope has a diameter of 20 mm.

Under test at a speed of 480 rev/min the brake load is 520 N. An indicator card taken during the test has an area of $2\,000$ mm^2 and length 60 mm. The indicator spring number is 15×10^3

$N/m^2/mm.$

Determine:

a) The indicated power.

b) The brake power.

c) The mechanical efficiency.

a) $\qquad P_M = \dfrac{Area}{Length} \times S \qquad\qquad L = 0{\cdot}3\,m$

$\qquad\qquad = \dfrac{2\,000}{60} \times 15 \times 10^3\ N/m^2$

$\qquad\qquad = 0{\cdot}5 \times 10^6\ N/m^2$

$\qquad A = \dfrac{\pi}{4} \times d^2 \qquad\qquad\qquad N = \dfrac{480}{60} \times \dfrac{1}{2}$

$\qquad\quad = \dfrac{\pi}{4} \times 0{\cdot}23^2 \qquad\qquad\qquad = 4 \text{ working strokes/sec.}$

$\qquad\quad = 0{\cdot}0415\,m^2$

$\qquad IP = P_M\,LAN$

$\qquad\qquad = 0{\cdot}5 \times 10^6 \times 0{\cdot}3 \times 0{\cdot}0415 \times 4\ W$

$\qquad\qquad = 24\,900\ W$

$\therefore\quad$ *Indicated power* $= 24{\cdot}9\,kW$

b) $\qquad n = \dfrac{480}{60} = 8\ rev/s;\ (W-S) = 520\,N;\ D = 1{\cdot}48\,m;$

$\qquad d = 0{\cdot}02\,m$

$\qquad BP = \pi n(W-S)(D+d)$

$\qquad\qquad = \pi \times 8 \times 520 \times 1{\cdot}5\ W$

$\qquad\qquad = 19\,600\ W$

$\therefore\quad$ *Brake power* $= 19{\cdot}6\,kW$

c) $\quad \eta\,\text{mech} = \dfrac{19{\cdot}6}{24{\cdot}9} = 0{\cdot}79\ or\ 79\%$

8 During a test on a twin cylinder two stroke engine, the following data was produced:

Mean effective pressure $0{\cdot}7 \times 10^6\ N/m^2$; swept volume of cylinder $300 \times 10^3\ mm^3$; engine speed 25 rev/s; brake load 107 N; effective brake radius 0·5 m.

Determine (a) IP, (b) BP, (c) η

a) $P_M = 0{\cdot}7 \times 10^6\ N/m^2$; $V = 300 \times 10^3 \times 10^{-9}\ m^3$; $N = 25$

$$IP = 2 \times P_M\,LAN \quad (2 \text{ cylinders})$$
$$= 2 \times P_M\,VN$$
$$= 2 \times 0{\cdot}7 \times 10^6 \times 300 \times 10^3 \times 10^{-9} \times 25 \text{ W}$$
$$= 10{\cdot}500 \text{ W}$$

$\therefore$ *Indicated power = 10·5 kW*

b)
$$BP = 2\pi n\,WL$$
$$= 2\pi \times 25 \times 107 \times 0{\cdot}5 \text{ W}$$

$\therefore$ *Brake power = 8·4 kW*

c)
$$\eta = \frac{BP}{IP}$$
$$= \frac{8{\cdot}4}{10{\cdot}5}$$
$$\eta\ mech = 0{\cdot}8$$

9 The mean effective pressure at 4 200 rev/min during a test on a 6 cylinder 4 stroke engine was $1{\cdot}1\,MN/m^2$. The cylinder bore was 80 mm and the piston stroke 68 mm. Determine:

a) The indicated power.

b) The brake power given an efficiency of 81%.

a) $P_M = 1{\cdot}1 \times 10^6\,N/m^2, \quad L = 68 \times 10^{-3} \text{ m}$

$A = \dfrac{\pi}{4} \times 80^2 \times 10^{-6}\,m^2 = 5\,030 \times 10^{-6}\,m^2$

$N = \dfrac{4\,200}{60} \times \dfrac{1}{2} = 35$ working strokes/second

$$IP = 6 \times P_M\,LAN \quad (6 \text{ cylinders})$$
$$= 6 \times 1{\cdot}1 \times 10^6 \times 68 \times 10^{-3} \times 5\,030 \times 10^{-6} \times 35 \text{ W}$$
$$= 79\,000 \text{ W}$$

$\therefore$ *Indicated power = 79 kW*

b)
$$\eta\ mech = \frac{BP}{IP}$$
$$BP = IP \times \eta$$
$$= 79 \times 0{\cdot}81 \text{ kW}$$

$\therefore$ *Brake power = 64 kW*

Examples

1 The indicator diagram of an engine has an area of $4\,000\,mm^2$ and a length of 80 mm. The spring number is $15 \times 10^3\,N/m^2/mm$. Determine the mean effective pressure of the engine.

2 An engine cylinder has a cross sectional area of 5 000 mm². The stroke of the piston is 90 mm. Determine the work done during one working stroke, given that the mean effective pressure is 0.8×10^6 N/m².

3 The mean effective pressure in an engine cylinder is 0.75 MN/m². If the bore is 0.15 m and the stroke is 0.18 m, find the work done during one cycle.

4 A single cylinder petrol engine having a piston of cross-sectional area 4 750 mm² and a stroke of 60 mm makes 25 working strokes per second. The mean effective pressure is 0.55 MN/m². Calculate the indicated power.

5 The piston of a single cylinder gas engine has a diameter of 200 mm and makes a stroke of 300 mm. Find the indicated power when the number of firing strokes/min is 140 and the mean effective pressure is 0.55×10^6 N/m².

6 A single cylinder two stroke engine is to be designed for an indicated power output of 7 kW at 3 000 rev/min. Calculate the required m.e.p. if the bore is 50.5 mm and the stroke 64.5 mm.

7 The mean effective pressure of a four cylinder, 4 stroke engine is 0.8 MN/m² at 5 400 rev/min. The stroke is 61 mm and the piston area is 4 750 mm². Find the indicated power.

8 On a test on a single cylinder two stroke engine, the speed was found to be 1 500 rev/min and the m.e.p. 0.65×10^6 N/m². Piston area: 2 400 mm². Piston stroke: 60 mm. Determine the indicated power.

9 An indicator card taken during a test on a 4 cylinder 4 stroke engine had an area of 4 050 mm² and a length of 75 mm. The spring number was 16.7×10^3 N/m²/mm. The engine which has a bore of 0.1 m and a stroke of 0.15 m was running at 8 rev/sec. Calculate:
a) The mean effective pressure.
b) The indicated power.

10 A twin cylinder 4 stroke engine develops maximum power at 5 500 rev/min. The swept volume of each cylinder is 240×10^3 mm³. Calculate the maximum indicated power if the mean effective pressure is 0.8 MN/m².

11 The following information is taken from a rope brake test on a gas engine:
diameter of flywheel: 1·1 m; diameter of rope: 20 mm; load (W): 220 N; spring balance reading (S): 38 N; engine speed: 750 rev/min. Determine the brake power.

12 A prony brake is used to test an electric motor. Calculate the brake power given the following details:
motor speed: 1 400 rev/min; length of arm: 0·3 m; weight on arm: 8 N.

13 A small water turbine when tested by means of a rope brake at 1 500 rev/min gave a brake power of 25 kW. Determine the spring balance reading if the effective flywheel diameter was 1·5 m and the mass of the load was 30 kg.

14 If the indicated power of the engine in example 11 is 9·5 kW, determine the mechanical efficiency of the engine.

15 A test on a single cylinder four stroke engine yielded the following results:
engine speed: 2 500 rev/min; mean effective pressure: 0·75 MN/m^2; stroke: 90 mm; piston area: $7·5 \times 10^3$ mm^2; effective flywheel diameter: 0·5 m; spring balance reading: 67 N; mass of the load: 19 kg. Calculate:
a) Indicated power.
b) Brake power.
c) $\eta\%$.

16 The test results for a 6 cylinder, 4 stroke engine are shown below:
Engine speed: 16 rev/s; indicator spring number: 20×10^3 $N/m^2/mm$; indicator diagram area: 2 100 mm^2; indicator diagram length: 50 mm; piston area: 7 000 mm^2; piston stroke: 90 mm; flywheel diameter: 0·775 m; rope diameter: 25 mm; Spring balance reading: 114 N; Mass of load: 65 kg. Determine:
a) The indicated power.
b) The brake power.
c) The mechanical efficiency.

17 A test on a steam engine was carried out using a prony brake. The brake arm had a length of 1·2 m and the mass of the load was 17·9 kg. Calculate:

a) The brake power, the engine speed being 250 rev/min.

b) The indicated power and friction power (power lost), if the mechanical efficiency of the engine is 0·85.

12
Equations of Motion

This chapter considers the motion of objects moving in a straight line with uniform motion.

Velocity is the rate of change of distance. The unit of velocity is the metre/second (m/s). It may also be measured in km/h, and care must be taken that the units in all equations are consistent.

$$\text{Average velocity} = \frac{\text{Total distance travelled}}{\text{Total time taken}}$$

or Distance travelled = Average velocity × time taken

Velocity-Time Graph

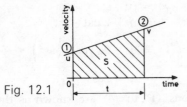

Fig. 12.1

Consider the graph shown where the velocity increases uniformly from ① to ② . Let u be the initial velocity, v be the final velocity after the time t, and s be the distance travelled in this time.

$$\text{Average velocity} = \frac{u+v}{2}$$

and: s = Average velocity × time

i.e. $\quad s = \left(\frac{u+v}{2}\right)t \quad . \quad . \quad . \quad . \quad . \quad$ ①

It should be noted that this is also the area under the velocity-time graph.

Acceleration is the rate of change of velocity. The unit is the metre/second squared (m/s^2) *i.e.* the change in velocity (m/s)/second (s).

$$\text{Average acceleration} = \frac{\text{Final velocity} - \text{initial velocity}}{\text{Time taken}}$$

NOTE: In Figure 12.1 the graph is a straight line and thus the acceleration is a constant.

$$i.e. \quad a = \frac{v - u}{t}$$

$$at = v - u$$

and thus:
$$v = u + at \quad . \quad . \quad . \quad . \quad . \quad ②$$

Substitute ($u + at$) for v in ①

$$s = \left(\frac{u + u + at}{2}\right)t$$

$$= \left(\frac{2u}{2} + \frac{at}{2}\right)t$$

$$i.e. \quad s = ut + \frac{1}{2}at^2 \quad . \quad . \quad . \quad . \quad ③$$

From ②,
$$v^2 = (u + at)^2$$
$$= u^2 + 2\,uat + a^2t^2$$
$$= u^2 + 2a\left(ut + \frac{1}{2}at^2\right)$$

$$i.e. \quad v^2 = u^2 + 2as \quad . \quad . \quad . \quad . \quad . \quad ④$$

The above equations ① to ④ are known as the *Equations of Motion* and will apply to any object moving with uniformly accelerated motion.

Worked Examples

1   A railway train accelerates from 30 km/h to 80 km/h in 8 s. Draw a velocity/time diagram and calculate:
   a) The acceleration.
   b) The distance travelled in this time.

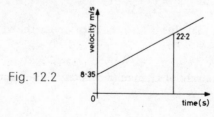

Fig. 12.2

$$u = 30\,km/h \qquad v = 80\,km/h \qquad\qquad t = 8\,s$$

$$= 30 \times \frac{10}{36}\,m/s \qquad = 80 \times \frac{10}{36}\,m/s$$

$$= 8\cdot35\,m/s \qquad\qquad = 22\cdot2\,m/s$$

$$v = u + at$$

$$22\cdot2 = 8\cdot35 + a \times 8$$

$$a = \frac{22\cdot2 - 8\cdot35}{8}$$

$$= \frac{13\cdot85}{8}$$

$\therefore$ *Acceleration of the train is $1\cdot73\,m/s^2$*

b) From the diagram, distance travelled = area under the graph

$$= \left(\frac{8\cdot35 + 22\cdot2}{2}\right) 8$$

$$= \frac{30\cdot55}{2} \times 8$$

i.e. *Distance travelled while accelerating is $122\cdot3\,m$.*

2   The driver of a car travelling at 60 km/h applies the brakes and brings the car to rest in 4 s. Calculate the acceleration of the car.

$$u = 60\,km/h \qquad v = 0 \qquad t = 4\,s$$

$$= 60 \times \frac{10}{36}\,m/s$$

$$= \frac{100}{6}\,m/s$$

$$v = u + at$$

$$0 = \frac{100}{6} + a \times 4$$

$$4a = -\frac{100}{6}\,m/s^2$$

$$a = -4\cdot16\,m/s^2$$

NOTE: The minus sign indicates that the car is slowing down: *i.e.* a retardation.

3   A motor car accelerates uniformly from rest to a speed of 36 km/h with an acceleration of $2\cdot5$ m/s^2. It then moves with uniform velocity for 20 s after which it is brought to rest with an acceleration of -2 m/s^2. Draw a velocity-time graph and determine:

a) The total time taken.
b) The total distance travelled.

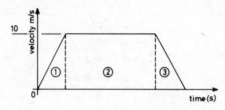

Fig. 12.3

a) PART 1

$$u = 0 \qquad v = 36 \text{ km/h} \qquad a = 2 \cdot 5 \text{ m/s}^2$$
$$\qquad\qquad\quad = 10 \text{ m/s}$$
$$v = u + at$$
$$10 = 0 + 2 \cdot 5 t_1$$
$$t_1 = \frac{10}{2 \cdot 5} = 4 \text{ s}$$

PART 2
$$t_2 = 20 \text{ s}$$

PART 3
$$v = 0 \qquad u = 10 \text{ m/s} \qquad a = -2 \text{ m/s}^2$$
$$v = u + at$$
$$0 = 10 - 2t_3$$
$$t_3 = 5 \text{ s}$$
$$\therefore \text{ Total time taken} = 4 + 20 + 5$$
$$= 29 \text{ s}$$

b) PART 1
$$s_1 = ut + \tfrac{1}{2}at_1{}^2$$
$$\quad = 0 + \tfrac{1}{2} \times 2 \cdot 5 \times 4^2$$
$$s_1 = 20 \text{ m}$$

PART 2
$$s_2 = ut_2$$
$$\quad = 10 \times 20$$
$$s_2 = 200 \text{ m}$$

PART 3
$$s_3 = ut_3 + \tfrac{1}{2}at_3{}^2$$
$$\quad = 10 \times 5 - \tfrac{1}{2} \times 2 \times 5^2$$
$$\quad = 50 - 25 \text{ m}$$
$$s_3 = 25 \text{ m}$$

$\therefore$ *Total distance travelled* $= 20 + 200 + 25$
$$= 245 \ m$$

4   A chute is used to slide bags of flour from a store to a lorry waiting below. The chute is 5 m long and it takes a bag 4 s to slide down it. Calculate:
   a) The acceleration.
   b) The final velocity of the bag.

   a) s $= 5$ m   u $= 0$ t $= 4$ s
      s $= ut + \frac{1}{2}at^2$
      $5 = 0 + \frac{1}{2}a4^2$
      a $= \dfrac{10}{16}$ m/s^2

   i.e. a $= 0 \cdot 625$ m/s^2

   b) v $= u + at$
      $= 0 + 0 \cdot 625 \times 4$
      v $= 2 \cdot 5$ m/s

5   A distress rocket is fired vertically into the air with an initial velocity of 49 m/s. Determine:
   a) The time taken to reach maximum height.
   b) The maximum height reached.
   c) The velocity of the rocket when at a height of 50 m.

   a) a $= -g = -9 \cdot 81$ m/s^2 u $= 49$ m/s   v $= 0$
      v $= u + at$
      $0 = 49 - 9 \cdot 81$ t
      t $= \dfrac{49}{9 \cdot 81}$

      $\therefore$ *Time taken to reach maximum height is 5 s.*

   b) s $= \left(\dfrac{u + v}{2}\right)t$

      $= \left(\dfrac{49 + 0}{2}\right) 5$

      $\therefore$ *Maximum height reached is 122·5 m.*

   c)   s $= 50$ m   u $= 49$ m/s   a $= -9 \cdot 81$ m/s^2
      $v^2 = u^2 + 2as$
      $v^2 = 49^2 - 2 \times 9 \cdot 81 \times 50$
      $v^2 = 2\,401 - 981$

$$v = \sqrt{1\,420} \text{ m/s}$$
$$\therefore \quad v = 37.7 \text{ m/s}$$

6 The tup of a pile driver accelerates at a rate of 9 m/s^2 from rest. It drops 4 m onto the pile. Determine:
   a) The velocity at the point of impact.
   b) The time taken for the stroke.

   a) $a = 9 \text{ m/s}^2$ $\qquad$ $s = 4 \text{ m}$ $\qquad$ $u = 0$
   $$v^2 = u^2 + 2as$$
   $$v^2 = 0 + 2 \times 9 \times 4 \text{ m/s}$$
   $$v = \sqrt{2 \times 9 \times 4} \text{ m/s}$$
   $$= 6\sqrt{2} \text{ m/s}$$
   $$\therefore \quad v = 8.5 \text{ m/s}$$

   b) $s = ut + \frac{1}{2}at^2$
   $$4 = 0 + \frac{1}{2} \times 9t^2$$
   $$t^2 = \frac{8}{9}$$
   $$t = \sqrt{\frac{8}{9}}$$
   $$\therefore \quad t = 0.94 \text{ s}$$

Examples

1 A car travels at a uniform speed of 72 km/h for 35 minutes. Find the total distance travelled.

2 Find the velocity after 6 s of an object which moves from rest with a uniform acceleration of 0.5 m/s.

3 A particle with an initial velocity of 5 m/s accelerates uniformly to a velocity of 10 m/s over a distance of 5 m. Determine the value of the acceleration.

4 During a performance test on a saloon car, the time taken to reach 90 km/h from rest was 15 s. Assuming uniform acceleration, calculate:
   a) The distance travelled in this time.
   b) The acceleration of the car.

5 A piston in a cylinder travels a distance of 8 cm from rest with uniform acceleration in a time of 8 ms. Calculate:
   a) The acceleration.
   b) The final velocity.

6 The speed of a VC 10 shortly after leaving the ground is 180 km/h. It accelerates uniformly to 900 km/h over a distance of 5 km. Determine:
   a) Its acceleration.
   b) The time taken to cover this distance.

7 A driver travelling at 108 km/h sees an obstacle ahead of him and applies the brakes. He comes to rest in 5 s from the time that he applies the brakes. Determine:
   a) The distance travelled in this time.
   b) The acceleration.

   NOTE: This is not the total distance travelled from the time the obstacle is sighted. It would take about 1·5 s between sighting the obstacle and applying the brakes at this speed.

8 A steam forging hammer has a velocity of 8·5 m/s when it hits the metal. It is brought to rest in 5 ms. Determine:
   a) The acceleration of the hammer.
   b) The amount that the metal is reduced by every blow: i.e. the distance required to bring the hammer to rest.

9 The cable of a tower crane breaks when the load is 45 m above the ground. Determine the time before the load hits the ground.

10 A car with a top speed of 126 km/h has a maximum acceleration of 3·5 m/s² and a maximum retardation of 5 m/s². Determine the shortest time required to cover 2·155 km. Draw a velocity/time graph for the process.

11 A racing car travels 540 m in 12 s. Calculate its final velocity and acceleration if the initial velocity is 10·8 km/h.

12 A mine hoist rises from the bottom of a mine 400 m deep with an acceleration of 0·8 m/s² for 10 s. It then continues with uniform velocity up to the surface. Draw a velocity/time graph and find:
   a) The uniform velocity.
   b) The distance that it rises while accelerating.
   c) The total time taken to rise to the surface.

13 A boy drops a ball from a window in a block of high flats. If it takes
3 s to reach the ground, determine:
  a) The distance that the ball drops.
  b) The number of flats up, each flat having a height of 3 m.Explain
     the fractional part of the answer.
  c) The velocity of the ball as it hits the ground.

14 A car starting from rest at a post accelerates uniformly and passes
the next post 225 m further on 15 s later. Find how long it takes
from starting until it reaches the third post 225 m from the second
and its velocity as it passes this third post.

13
Strength of Materials (1)
Direct Stress and Strain

Tension   Materials (such as chains, wires, thin rods and various steel sections) when subjected to an external axial pull are said to be in *tension* and the material is in a state of *tensile stress*.

Tensile Stress σ is the internal reactive force set up within a material resisting an axial pull.

Tensile Strain ε is the elongation or stretch of a material when it is in a state of *tensile stress*.

Tie   In a structure, a member subjected to a *tensile force* is called a *tie*.

Compression   Materials (forming columns, supports etc.) when subjected to an external axial thrust are said to be in a state of *compressive stress*.

Compressive Stress σ is the internal reactive force set up within a material resisting an axial thrust.

Compressive Strain ε is the contraction or shortening of a material when it is in a state of *compressive stress*.

Strut   In a structure, a member subjected to a compressive force is called a *strut*.

Load is the force, or resultant of a system of forces applied to a body (*i.e.* an external force).

Stress   An external force acting on a solid material produces opposing resistances within the material. The combined effect of these resistances, considered over a sectional area of the material is called the *intensity of stress* or *stress*.

Direct Stress   Direct stresses occur when the cross sectional area being stressed is at right angles to the external force (Figure 13.1). The direction of the external force determines the nature of the stress in the material, *e.g.*

An axial pull produces a *Tensile Stress*.

An axial thrust produces a *Compressive Stress*.

Tensile and compressive stresses are therefore *Simple Direct Stresses*.

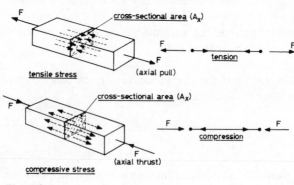

Fig. 13.1

Measurement of Direct Stress (Tensile and Compressive)

This is the internal reaction per unit area of cross-section and is expressed in N/m^2.

NOTE: For large forces (loads) kN/m^2 or MN/m^2 may be used.

$$\text{Stress} = \frac{\text{Load (N)}}{\text{Area (m}^2)}$$

$$\sigma = \frac{F}{A_X} \; N/m^2$$

(σ is the greek letter 'sigma'.)

F = applied direct load
A_X = cross-sectional area of the material
σ = direct stress

Before considering worked examples attention is drawn to some metric equivalents which are often used.

$$1\,\text{m} = 1\,\text{mm} \times 10^3 \qquad\qquad 1\,\text{kN/m}^2 = 1\,\text{N/m}^2 \times 10^3$$
$$1\,\text{mm} = 1\,\text{m} \times 10^{-3} \qquad\qquad 1\,\text{MN/m}^2 = 1\,\text{N/m}^2 \times 10^6$$
$$1\,\text{m}^2 = 1\,\text{mm}^2 \times 10^6 \qquad\qquad 1\,\text{GN/m}^2 = 1\,\text{N/m}^2 \times 10^9$$
$$1\text{mm}^2 = 1\,\text{m}^2 \times 10^{-6} \qquad\qquad 1\,\text{MN/m}^2 = 1\,\text{N/mm}^2$$

Worked Examples

1 A solid circular bar 40 mm diameter is subjected to a compressive load of 90 kN. Calculate the intensity of compressive stress.

Cross-sectional area of the bar

$$A_X = \frac{\pi}{4}\,d^2$$

$$= \frac{\pi}{4} \times 40 \times 40 \text{ mm}^2$$

$$= 400\pi \text{ mm}^2$$

$$= 1\,256 \text{ mm}^2 \qquad \text{Fig. 13.2}$$

Compressive stress $= \dfrac{\text{Load}}{\text{Cross-sectional area}}$

$$\sigma = \frac{F}{A_X}$$

$$= \frac{90 \times 10^3}{1\,256} \text{ N/mm}^2$$

$$= \frac{90\,000}{1\,256} \text{ N/mm}^2$$

$$= 71 \cdot 6 \text{ N/mm}^2$$

$$= 71 \cdot 6 \text{ MN/m}^2$$

The compressive stress in the bar $= 71 \cdot 6$ MN/m^2

NOTE:

$$71 \cdot 6\,\frac{\text{N}}{\text{mm}^2} = 71 \cdot 6\,\frac{\text{N}}{\text{m}^2 \times 10^{-6}}$$

$$= 71 \cdot 6 \times 10^6 \text{ N/m}^2$$

$$= 71 \cdot 6 \text{ MN/m}^2$$

2 A mass of 800 kg hangs from a solid vertical bar of rectangular cross-section 30 mm × 20 mm. Calculate the tensile stress in the material.

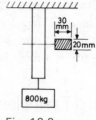

Cross-sectional area of the bar

$$A_X = 30 \times 20 \text{ mm}^2$$
$$= 600 \text{ mm}^2$$

Fig. 13.3

Gravitational force acting on the bar

$$F = \text{mass} \times g \text{ N} \qquad (g = 9 \cdot 81 \text{ m/s}^2)$$
$$= 800 \times 9 \cdot 81 \text{ N}$$
$$= 7\,850 \text{ N}$$

Tensile stress $= \dfrac{F}{A_X}$

$$\sigma = \frac{7\,850}{600} \text{ N/mm}^2$$
$$= 13 \cdot 1 \text{ N/mm}^2$$
$$= 13 \cdot 1 \text{ MN/m}^2$$

The tensile stress in the bar $= 13 \cdot 1 \text{ MN/m}^2$

3   A copper tube 20 mm outside diameter and 14 mm inside diameter supports a mass of 340 kg as shown. Calculate the direct stress in the tube material.

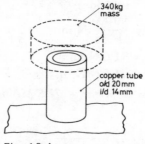

Fig. 13.4

Let 'D' be the outside diameter of the tube.
Let 'd' be the inside diameter of the tube.
Area of cross section (*i.e.* the wall thickness)

$$A_X = \left(\frac{\pi}{4} D^2 - \frac{\pi}{4} d^2\right) \text{ mm}^2$$

Gravitational force
$$F = mg$$

$$A_X = \left(\frac{\pi}{4} \times 20^2 - \frac{\pi}{4} \times 14^2\right) mm^2 \qquad F = 340 \times 9 \cdot 81 \text{ N}$$
$$= (100\pi - 49\pi) \text{ mm}^2 \qquad\qquad = 3\,330 \text{ N}$$
$$= 51\pi \text{ mm}^2$$
$$= 160 \text{ mm}^2$$

$$\text{Compressive stress} = \frac{F}{A_X}$$
$$= \frac{3\,330}{160} \text{ N/mm}^2$$
$$= 20 \cdot 8 \text{ N/mm}^2$$
$$= 20 \cdot 8 \text{ MN/m}^2$$

The direct (compressive) stress = 20·8 MN/m²

4   An assembled circular tie-rod designed to a diameter of 20 mm and to withstand a tensile load of 10 kN, was found to measure 18 mm diameter. Calculate the increase in stress for the same designed force.

To find the designed stress:
Designed cross-sectional area of the rod

$$A_X = \frac{\pi}{4} d^2$$
$$= \frac{\pi}{4} \times 20^2 \text{ mm}^2$$
$$= 100\pi \text{ mm}^2$$
$$= 314 \text{ mm}^2$$

$$\text{Designed stress} = \frac{F}{A_X}$$
$$= \frac{10\,000}{314} \text{ N/mm}^2$$
$$= 31 \cdot 8 \text{ N/mm}^2$$
$$= 31 \cdot 8 \text{ MN/m}^2$$

To find the actual stress:
Actual cross-sectional area of the rod

$$A_X = \frac{\pi}{4} d^2$$
$$= \frac{\pi}{4} \times 18^2 \text{ mm}^2$$

$$= 81\pi \text{ mm}^2$$
$$= 254 \text{ mm}^2$$

Actual stress $= \dfrac{F}{A_X}$

$$= \dfrac{10\,000}{254} \text{ N/mm}^2$$

$$= 39\cdot4 \text{ N/mm}^2$$

$$= 39\cdot4 \text{ MN/m}^2$$

Increase in stress $=$ Actual stress $-$ designed stress
$$= (39\cdot4 - 31\cdot8) \text{ MN/m}^2$$
$$= 7\cdot6 \text{ MN/m}^2$$

Increase in stress $= 7\cdot6 \text{ MN/m}^2$

5   The end arrangement of a tie-bar is shown. The shank (16 mm diameter) is subjected to a tensile force F. If the tensile stress in the shank is equal to the compressive stress in the head, calculate diameter 'D' of the head.

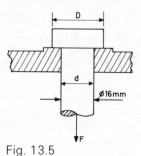

Fig. 13.5

Tensile stress in the shank $= \dfrac{F}{A_X}$

$$= \dfrac{F}{\dfrac{\pi}{4}d^2} \text{ N/mm}^2$$

$$= \dfrac{4F}{\pi d^2} \text{ N/mm}^2$$

To find the compressive stress in the head:

NOTE: Cross-sectional area of the head under compression

$$= \dfrac{\pi}{4}D^2 - \dfrac{\pi}{4}d^2$$

$$= \frac{\pi}{4}(D^2 - d^2)$$

Compressive stress in the head $= \dfrac{F}{A_X}$

$$= \frac{F}{\frac{\pi}{4}(D^2 - d^2)}$$

$$= \frac{4F}{\pi(D^2 - d^2)}$$

Given that:

Compressive stress in head = Tensile stress in shank

$$\frac{4F}{\pi(D^2 - d^2)} = \frac{4F}{\pi d^2}$$

$$D^2 - d^2 = d^2$$

$$D^2 = 2d^2$$

$$D^2 = 2 \times 16^2 \text{ mm}^2$$

$$D^2 = 512 \text{ mm}^2$$

$$D = 22 \cdot 6 \text{ mm}$$

$\therefore$ *Diameter of the head = 22·6 mm*

Strain is the change of form or dimensions of a solid due to the material being in a state of stress.

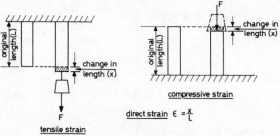

Fig. 13.6

Measurement of Direct Strain (Tensile and Compressive)

This is the change of length per unit of the original length.

$$\text{Strain} = \frac{\text{Change in length}}{\text{Original length}}$$

$$\varepsilon = \frac{x}{L}$$

x = Change in length
L = Original length
ε = Direct strain

(ε is the Greek letter 'epsilon'.)

NOTE: Strain has no units.

Worked Examples

1   Due to a tensile load, a bar 2 m long extends 1·5 mm. Calculate the tensile strain.

$$\text{Tensile strain} = \frac{\text{Change in length}}{\text{Original length}}$$

$$\varepsilon = \frac{x}{L}$$

$$= \frac{1\cdot5 \text{ mm}}{2 \text{ m}}$$

$$= \frac{1\cdot5}{2 \times 10^3}$$

$$\therefore \text{ Tensile strain} = 0\cdot75 \times 10^{-3}$$

2   The fractional strain in a rod is $0\cdot6 \times 10^{-3}$. How much will a rod 2 m in length stretch under tensile stress?

$$\text{Tensile strain} = \frac{\text{Change in length}}{\text{Original length}}$$

$$\varepsilon = \frac{x}{L}$$

$$x = \varepsilon \times L$$

$$= 0\cdot6 \times 10^{-3} \times 2 \times 10^3 \text{ mm}$$

$$= 1\cdot2 \text{ mm}$$

The extension of the rod $= 1\cdot2$ *mm*

3   A compressive load causes a bar to shorten by 1·5 mm. Calculate the original length of the bar if the strain is $0\cdot5 \times 10^{-3}$.

$$\text{Compressive strain} = \frac{\text{Change in length}}{\text{Original length}}$$

$$\varepsilon = \frac{x}{L}$$

$$L = \frac{x}{\varepsilon}$$

$$= \frac{1 \cdot 5 \text{ mm}}{0 \cdot 5 \times 10^{-3}}$$

$$= 3 \times 10^3 \text{ mm}$$

$$= 3\,000 \text{ mm}$$

$$= 3 \text{ m}$$

The original length of the bar $= 3 m$

Hooke's Law   Robert Hooke (1635–1702) carried out a series of experiments on various materials and introduced his Theory of Elasticity.

When materials remain fully elastic (not stretched beyond a certain limit):

a) they return to their original shape when the external force (load) is removed;

b) the extension is directly proportional to the load, *i.e.*

If a 1 N Force causes an extension of 'x' mm

then a 2 N Force causes an extension of '2x' mm

and a 3 N Force causes an extension of '3x' mm

Modulus of Elasticity (Young's Modulus)   The ratio of stress set up in a material and the strain produced within the proportional limit is constant for that material. The constant 'E' is called 'Modulus of Elasticity' or 'Young's Modulus'. Thomas Young (1773–1829) used the Stress/Strain ratio in experiments for tensile loadings and the name 'Modulus of Elasticity' has been attributed to him. Although compressive loading causes a slight variation in the Stress/Strain ratio from that of tensile loading, 'E' is normally taken as the same for both.

Young's Modulus $= \dfrac{\text{Stress}}{\text{Strain}} \text{ N/m}^2$

$$E = \frac{\sigma}{\varepsilon}$$

σ = Direct Stress

ε = Direct Strain

E = Young's Modulus of Elasticity

Substituting the symbols for direct stress and direct strain in the above formula:

$$E = \frac{\sigma}{\varepsilon}$$

$$E = \frac{\dfrac{F}{A_X}}{\dfrac{x}{L}}$$

$$E = \frac{F \times L}{A_X \times x}$$

i.e.

Young's Modulus of Elasticity $= \dfrac{\text{Load} \times \text{Original length}}{\text{Area}_X \times \text{Change in length}}$ N/m^2

Worked Examples

1 A mild steel bar 3·5 m long and uniform cross-section 30 mm × 15 mm is subjected to a tensile load of 30 kN. If the strain for this condition is $0·33 \times 10^{-3}$ calculate the Modulus of Elasticity for the bar.

To find the tensile stress:
Cross-sectional area of the bar

$$A_X = 30 \times 15 \, \text{mm}^2$$
$$= 450 \, \text{mm}^2$$
$$= 450 \times 10^{-6} \, \text{m}^2$$

$$\text{Tensile stress} = \frac{F}{A_X}$$

$$= \frac{30 \times 10^3}{450 \times 10^{-6}} \, \text{N/m}^2$$

$$= 0·0666 \times 10^9 \, \text{N/m}^2$$
$$= 66·6 \times 10^6 \, \text{N/m}^2$$
$$= 66·6 \, \text{MN/m}^2$$

$$\text{Tensile strain} = 0·33 \times 10^{-3}$$

$$\text{Young's Modulus (E)} = \frac{\text{Stress}(\sigma)}{\text{Strain}(\varepsilon)}$$

$$E = \frac{66·6 \times 10^6}{0·33 \times 10^{-3}} \, \text{N/m}^2$$

$$E = 202 \times 10^9 \, \text{N/m}^2$$
$$E = 202 \, \text{GN/m}^2$$

Young's Modulus of Elasticity for the mild steel bar = 202 GN/m².

2 Four vertical aluminium 'T' bars, each 800 mm in length and 50 mm × 50 mm × 8 mm, are used to support a fresh water tank on board a motor yacht. If the mass of the uniform tank, when full, is 3 tonne and each support shortens 0·12 mm calculate for this condition (for each support):
a) The compressive stress
b) The compressive strain
c) Young's Modulus of Elasticity for aluminium.

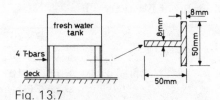

Fig. 13.7

Gravitational force = mass × g
$$= 3\,000 \times 9·81 \, \text{N}$$
$$= 29\,430 \, \text{N}$$
$$= 29·43 \, \text{kN}$$

This force is supported by four 'T' bars.
Force acting on one 'T' bar

$$F = \frac{29·43 \, \text{kN}}{4}$$

$$= 7·36 \, \text{kN}$$

The cross-sectional area of one 'T' bar support
$$A_X = [(50 \times 8) + (42 \times 8)] \, \text{mm}^2$$
$$= (400 + 336) \, \text{mm}^2$$
$$= 736 \, \text{mm}^2$$

a) Compressive stress $\sigma = \dfrac{F}{A_X}$

$$= \frac{7·36 \times 10^3}{736} \, \text{N/mm}^2$$

$$= \frac{7\,360}{736} \, \text{N/mm}^2$$

$$= 10 \, \text{N/mm}^2$$
$$= 10 \, \text{MN/m}^2$$

The compressive stress in one support $= \mathit{10 \, MN/m^2}$

b) Compressive strain $\varepsilon = \dfrac{\text{x}}{\text{L}}$

$$= \frac{0 \cdot 12}{800}$$
$$= 0 \cdot 00015$$
$$= 0 \cdot 15 \times 10^{-3}$$

The compressive strain in one support $= \mathit{0 \cdot 15 \times 10^{-3}}$

c) Young's Modulus of Elasticity $\text{E} = \dfrac{\sigma}{\varepsilon}$

$$= \frac{10 \times 10^6}{0 \cdot 15 \times 10^{-3}} \, \text{N/m}^2$$
$$= 66 \cdot 7 \times 10^9 \, \text{N/m}^2$$
$$= 66 \cdot 7 \, \text{GN/m}^2$$

Young's Modulus of Elasticity for aluminium $= \mathit{66 \cdot 7 \, GN/m^2}.$

3   A 2 m length of tubular scaffolding, outside diameter 100 mm, inside diameter 80 mm is subjected to a compressive stress of 15 MN/m². Calculate:
  a) The compressive load.
  b) The amount that the scaffolding shortens.
     Take E = 200 GN/m².

  a) To find the force acting on the scaffolding.
     Cross-sectional area of the tube

$$\text{A}_\text{X} = \frac{\pi}{4}\text{D}^2 - \frac{\pi}{4}\text{d}^2 \qquad \text{D = outside diameter}$$

$$\text{d = inside diameter}$$

$$= (2\,500\pi - 1\,600\pi) \, \text{mm}^2$$
$$= 900\pi \, \text{mm}^2$$
$$= 2\,826 \times 10^{-6} \, \text{m}^2$$

Using stress formula

$$\sigma = \frac{\text{F}}{\text{A}_\text{X}}$$
$$\text{F} = \sigma \times \text{A}_\text{X}$$

$$= 15 \times 10^6 \times 2\,826 \times 10^{-6}\,\text{N}$$
$$= 42\,400\,\text{N}$$
$$= 42 \cdot 4\,\text{kN}$$

The compressive load $= 42 \cdot 4\,kN$

b) Young's Modulus $= \dfrac{\text{Stress}}{\text{Strain}}$

$$\text{E} = \frac{\sigma}{\varepsilon}$$

$$\varepsilon = \frac{\sigma}{\text{E}}$$

$$\varepsilon = \frac{15 \times 10^6}{200 \times 10^9}$$

Strain $(\varepsilon) = 0 \cdot 075 \times 10^{-3}$

Using strain formula

$$\varepsilon = \frac{\text{x}}{\text{L}}$$

$$\text{x} = \varepsilon \times \text{L}$$
$$= 0 \cdot 075 \times 10^{-3} \times 2 \times 10^3\,\text{mm}$$
$$= 0 \cdot 15\,\text{mm}$$

The tubular piece of scaffolding shortens 0·15 mm

Tensile Test (Mild Steel)   Before materials are regarded as suitable for specific engineering purposes a sample of the metal to be used (in the form of a test piece made to a standard shape) is mounted on a specially designed machine which gradually exerts an increasing tensile load until eventually the test piece fractures. Throughout the test recordings are made and a Load/Extension graph plotted.

Figure 13.8 illustrates the change in shape of the test piece during the test.

Percentage Elongation $= \dfrac{\text{Final Length} - \text{Original Length}}{\text{Original Length}} \times 100$

% Elongation $= \dfrac{\text{L}_\text{F} - \text{L}_0}{\text{L}_0} \times 100$

Percentage Reduction in Area $= \dfrac{\text{Original Area} - \text{Final Area}}{\text{Original Area}} \times 100$

$$\% \text{ Reduction } A_X = \frac{A_0 - A_F}{A_0} \times 100$$

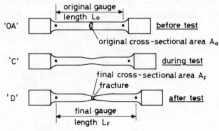

Fig. 13.8

Figure 13.9a shows a typical Load/Extension graph for a mild steel (ductile material) test piece.

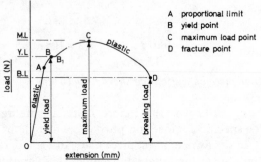

Fig. 13.9a

Observations from the Load/Extension graph (Figure 13.9a):

'OA': *Proportional stage* ... straight line ... extension proportional to load ... elastic.

'A': *Proportional limit.*

'AB': Slightly more increase in extension ... not proportional ... still elastic.

'B': *Yield point.*

'BC': *Plastic stage 1* ... general extension ... permanent set.

NOTE: Immediately after point 'B' (yield point) the material suddenly gives and from 'B' to 'B₁' there is a rapid extension for little or no additional load. At point 'B₁' the metal recovers some of its resistance and a further increase in applied load causes further extension.

'C': *Maximum load* . . . test piece can withstand no further load . . . a 'waist' or 'neck' forms in the test piece.

'CD': *Plastic stage 2* . . . local distortion . . . decrease in load.

'D': *Fracture* occurs.

Figure 13.9b: Stress/Strain diagrams are exactly the same shape as Load/Extension diagrams since the cross-sectional area is considered constant for all stress calculations.

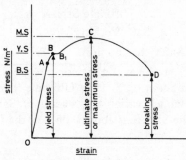

Fig. 13.9b

From the tensile test the Yield Stress and Ultimate (maximum) Stress are obtained.

Yield Stress is the stress at the point (yield point) where the strain suddenly increases and the material receives a 'permanent set'.

$$\text{Yield Stress} = \frac{\text{Yield Load}}{\text{Original Area}} N/m^2 \quad Y.S. = \frac{Y.L.}{A_X}$$

Ultimate Stress (Tensile and Compressive) is the maximum stress in the material just before fracture.

$$\text{Ultimate Stress} = \frac{\text{Maximum Load}}{\text{Original Area}} N/m^2 \quad \left.\begin{array}{l} U.T.S. \\ U.C.S. \end{array}\right\} = \frac{M.L.}{A_X}$$

Safe Working Stress is a selected stress to which a material is subjected in practice. It must not exceed the stress at the proportional limit for that material and is usually a simple fraction of the ultimate stress.

Factor of Safety (F.O.S.) is the ratio of the Maximum Load to Safe Working Load or Ultimate Stress to Safe Working Stress. The Factor of Safety value is selected such that the Safe Working Stress will not be greater than the stress at the proportional limit. This allows a margin of safety for unforeseen happenings in a material or structure.

$$\text{Factor of Safety} = \frac{\text{Maximum Load}}{\text{Safe Working Load}} \qquad \text{F.O.S.} = \frac{\text{M.L.}}{\text{S.W.L.}}$$

and assuming the original cross-sectional area constant:

$$\text{Factor of Safety} = \frac{\text{Ultimate Stress}}{\text{Safe Working Stress}} \qquad \text{F.O.S.} = \frac{\text{U.S.}}{\text{S.W.S.}}$$

NOTE: Factor of Safety has no units.

For 'dead loads' (stationary) a Factor of Safety of 4 is normal; *i.e.* the S.W.S. is about 0·25 of the Ultimate Tensile Stress.

For 'live loads' (shock variable) a Factor of Safety of up to 16 is common. *i.e.* the S.W.S. is about 0·0625 of the Ultimate Tensile Stress.

Abbreviations used in worked examples:

Safe Working Load. . . . . S.W.L.

Yield Load . . . . Y.L.

Maximum Load . . . . M.L.

Breaking Load . . . . .B.L.

Original Area of cross section . . . . A_X

Area of cross section at fracture . . . A_F

Safe Working Stress . . . . . . S.W.S. $= \dfrac{\text{S.W.L.}}{A_X} \text{N/m}^2$

Yield Stress . . . . Y.S. $= \dfrac{\text{Y.L.}}{A_X} \text{N/m}^2$

Ultimate Tensile Stress . . . . U.T.S. $= \dfrac{\text{M.L.}}{A_X} \text{N/m}^2$

Ultimate Compressive Stress . . U.C.S. $= \dfrac{\text{M.L.}}{A_X} \text{N/m}^2$

Breaking Stress (nominal). . . . . B.S. $= \dfrac{\text{B.L.}}{A_X} \text{N/m}^2$

Factor of Safety . . . . .F.O.S. $= \dfrac{\text{U.T.S.}}{\text{S.W.S.}} \text{ or } \dfrac{\text{U.C.S.}}{\text{S.W.S.}}$

Worked Examples

1 A mild steel rod 20 mm diameter fractures under a tensile load of 140 kN; calculate:

a) The ultimate tensile stress.

If a similar piece of steel carries a safe working load of 35 kN calculate:

b) The factor of safety.

c) The safe working stress.

a) The cross-sectional area of the rod

$$A_X = \frac{\pi}{4}d^2$$

$$= 100\pi \text{ mm}^2$$

$$= 314 \times 10^{-6} \text{ m}^2$$

$$\text{Ultimate tensile stress} = \frac{M.L.}{A_X}$$

$$= \frac{140 \times 10^3}{314 \times 10^{-6}} \text{ N/m}^2$$

$$= 0\cdot446 \times 10^9 \text{ N/m}^2$$

$$= 446 \text{ MN/m}^2$$

The ultimate tensile stress $= 446 \text{ MN/m}^2$

b)

$$\text{Factor of safety} = \frac{\text{Maximum load}}{\text{Safe working load}}$$

$$= \frac{140}{35}$$

$$= 4$$

The factor of safety $= 4$

c)

$$\text{Factor of safety} = \frac{\text{U.T.S.}}{\text{S.W.S.}}$$

$$\text{S.W.S.} = \frac{\text{U.T.S.}}{\text{F.O.S.}}$$

$$= \frac{446}{4} \text{ MN/m}^2$$

$$= 111\cdot5 \text{ MN/m}^2$$

The safe working stress $= 111\cdot5 \text{ MN/m}^2$

2 Figure 13.10 shows a steel bar ① holding a metal canopy in position. The bar has a rectangular cross-section 70 mm × 20 mm

and is 4·5 m in length. The maximum S.W.L. allows the bar to extend 1·5 mm. If the F.O.S. is 6 and Young's Modulus for steel taken as 206 GN/m², calculate:

a) The tensile load acting on the bar.
b) The U.T.S. of the steel bar.

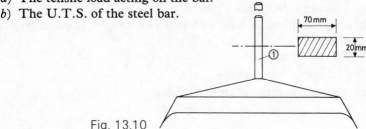

Fig. 13.10

The bar is in tension.

a) Tensile strain

$$\varepsilon = \frac{x}{L} \qquad\qquad x = 1.5 \text{ mm}$$
$$= \frac{1.5}{4.5 \times 10^3} \qquad L = 4.5 \times 10^3 \text{ mm}$$
$$= 0.33 \times 10^{-3}$$

Young's Modulus

$$E = \frac{\sigma}{\varepsilon}$$
$$\sigma = E \times \varepsilon$$
$$= 206 \times 10^9 \times 0.33 \times 10^{-3} \text{ N/m}^2$$
$$= 206 \times 0.33 \times 10^6 \text{ N/m}^2$$
$$= 68 \times 10^6 \text{ N/m}^2$$
$$\text{S.W.S.} = 68 \text{ MN/m}^2$$

Cross-sectional area of bar

$$A_X = 70 \times 20 \text{ mm}^2$$
$$= 1\,400 \text{ mm}^2$$
$$= 1\,400 \times 10^{-6} \text{ m}^2$$

Tensile stress (S.W.S.)

$$\sigma = \frac{F}{A_X}$$
$$F = \sigma \times A_X$$
$$= 68 \times 10^6 \times 1\,400 \times 10^{-6} \text{ N}$$
$$= 95\,200 \text{ N}$$

$$= 95 \cdot 2 \text{ kN}$$

The tensile load acting on the bar = 95·2 kN

b)   Factor of safety

$$\text{F.O.S.} = \frac{\text{U.T.S.}}{\text{S.W.S.}}$$

$$\text{U.T.S.} = \text{F.O.S.} \times \text{S.W.S.}$$
$$= 6 \times 68 \text{ MN/m}^2$$
$$= 408 \text{ MN/m}^2$$

The ultimate tensile stress of the steel bar = 408 MN/m²

3   In a structure a tie-rod is designed to carry a load of 170 kN. The rod is 3·5 m long, the U.T.S. is 430 MN/m² and F.O.S. is 6. If the Modulus of Elasticity for the material is 200 GN/m² calculate:
a) The safe working stress.
b) The diameter of the rod to the nearest millimetre.
c) The extension of the rod.

a) Factor of safety

$$\text{F.O.S.} = \frac{\text{U.T.S.}}{\text{S.W.S.}}$$

$$\text{S.W.S.} = \frac{\text{U.T.S.}}{\text{F.O.S.}}$$

$$= \frac{430}{6} \text{ MN/m}^2$$

$$= 71 \cdot 67 \text{ MN/m}^2$$

The safe working stress = 71·67 MN/m²

b) $$\text{S.W.S.} = \frac{\text{F}}{\text{A}_X}$$

$$\text{A}_X = \frac{\text{F}}{\text{S.W.S.}}$$

$$= \frac{170 \times 10^3}{71 \cdot 67 \times 10^6} \text{ m}^2$$

$$= 2 \cdot 37 \times 10^{-3} \text{ m}^2$$

$$= 2 \cdot 37 \times 10^{-3} \times 10^6 \text{ mm}^2$$

$$\frac{\pi}{4} \text{d}^2 = 2\,370 \text{ mm}^2$$

$$\text{d}^2 = \frac{2\,370 \times 4}{\pi} \text{ mm}^2$$

$$d^2 = 3\ 020\ \text{mm}^2$$
$$d = 55\ \text{mm}$$

The diameter of the rod = 55 mm

c) To find the strain

$$E = \frac{\sigma}{\varepsilon} \qquad \sigma = \text{S.W.S.}$$

$$\varepsilon = \frac{\text{S.W.S.}}{E}$$
$$= \frac{71 \cdot 67 \times 10^6}{200 \times 10^9}$$
$$= 0 \cdot 358 \times 10^{-3}$$

Tensile strain

$$\varepsilon = \frac{x}{L}$$
$$x = L \times \varepsilon$$
$$= 3 \cdot 5 \times 10^3 \times 0 \cdot 358 \times 10^{-3}\ \text{mm}$$
$$= 1 \cdot 25\ \text{mm}$$

The extension of the rod = 1·25 mm

4 The results of a tensile test on a mild steel test piece are given below. Draw the Load/Extension graph and from the graph take readings to calculate the Modulus of Elasticity for mild steel.

Gauge length of specimen 56·5 mm; diameter 11·3 mm;

Load (kN)	1	2	3	4	5
Extension (mm × 10^{-3})	2·8	5·7	8·5	11·4	14·1

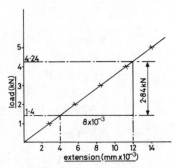

Fig. 13.11

Cross-sectional area of the test piece

$$A_X = \frac{\pi}{4} d^2$$

$$= \frac{\pi}{4} \times 11 \cdot 3 \times 11 \cdot 3 \text{ mm}^2$$

$$= 100 \text{ mm}^2$$

From the graph . . . Load (L)   = 2·84 kN

Extension (x) = 8×10^{-3} mm

$$\text{Tensile stress } \sigma = \frac{F}{A_X}$$

$$= \frac{2 \cdot 84 \times 10^3}{100} \text{ N/mm}^2$$

$$= 28 \cdot 4 \text{ N/mm}^2$$

$$= 28 \cdot 4 \text{ MN/m}^2$$

$$\text{Tensile strain } \varepsilon = \frac{x}{L}$$

$$= \frac{8 \times 10^{-3}}{56 \cdot 5}$$

$$= 0 \cdot 142 \times 10^{-3}$$

$$\text{Young's Modulus (E)} = \frac{\text{Stress } (\sigma)}{\text{Strain } (\varepsilon)}$$

$$= \frac{28 \cdot 4 \times 10^6}{0 \cdot 142 \times 10^{-3}} \text{ N/m}^2$$

$$= 200 \times 10^9 \text{ N/m}^2$$

$$= 200 \text{ GN/m}^2$$

Young's Modulus for the mild steel test piece = 200 GN/m²

5   The following results were obtained from a tensile test on a mild steel test piece.

Diameter of test piece . . . . 11·3 mm
Gauge length . . . . 56·5 mm
Diameter after fracture . . . . 8·1 mm
Length after fracture . . . . 65·1 mm
Load at yield point . . . . 25 kN
Maximum load . . . . 48 kN

Calculate:

a) The ultimate tensile stress.

b) The yield stress.

c) The percentage elongation.

d) The percentage reduction in area.

To find the cross-sectional area of test piece

$$A_X = \frac{\pi}{4} d^2$$

$$= \frac{\pi \times 11 \cdot 3^2}{4} \text{ mm}^2$$

$$= 100 \text{ mm}^2$$

a)   U.T.S. $= \dfrac{\text{M.L.}}{A_X}$

$$= \frac{48\,000}{100} \text{ N/mm}^2$$

$$= 480 \text{ N/mm}^2$$

$$= 480 \text{ MN/m}^2$$

The ultimate tensile stress = 480 MN/m²

b)   Y.S. $= \dfrac{\text{Y.L.}}{A_X}$

$$= \frac{25\,000}{100} \text{ N/mm}^2$$

$$= 250 \text{ N/mm}^2$$

$$= 250 \text{ MN/m}^2$$

The yield stress = 250 MN/m²

c)   % Elongation $= \dfrac{L_F - L_O}{L_O} \times 100$

$$= \frac{65 \cdot 1 - 56 \cdot 5}{56 \cdot 5} \times 100$$

$$= \frac{8 \cdot 6}{56 \cdot 5} \times 100$$

$$= 0 \cdot 152 \times 100$$

$$= 15 \cdot 2\%$$

The percentage elongation = 15·2%

d)   Area after fracture $= \dfrac{\pi}{4} d^2$

$$A_F = \frac{\pi \times 8 \cdot 1 \times 8 \cdot 1}{4} \text{ mm}^2$$

$$= 51 \cdot 5 \text{ mm}^2$$

$$\% \text{ Reduction } A_X = \frac{A_O - A_F}{A_O} \times 100$$

$$= \frac{100 - 51 \cdot 5}{100} \times 100$$

$$= 48 \cdot 5 \%$$

The percentage reduction in area $= 48 \cdot 5\%$

Examples

1 Metal bars of various cross-sections are subjected to tensile and compressive loads as shown (Figure 13.12). Calculate:
   a) The direct stress in bars (a) and (b). Name the nature of the stress in each case.
   b) The diameter of bar (c) if the tensile stress in the bar is 50 MN/m².
   c) The rectangular dimensions of bar (d) if the ratio of L:B is 5:3 and the tensile stress 20·45 MN/m².

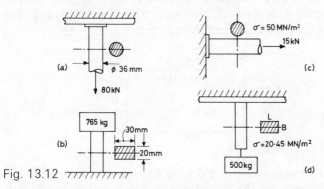

Fig. 13.12

2 The compressive stress set up in a steel bar of square cross section (60 mm side) is 40 MN/m². What compressive load is applied?

3 During a circus act an acrobat holds on to a 8 mm diameter rope which hangs vertically from a beam. If the stress in the rope is 15·6 MN/m² calculate the mass of the acrobat.

4 In a ship's engine room a platform providing access to engine parts is suspended from deck beams by means of circular steel rods. Figure 13.13 shows such a rod (30 mm diameter, with square head 60 mm side) which is designed to take a safe working load of 60 kN.

Calculate for this load:

a) The tensile stress in each rod.

b) The compressive stress in the head of each rod.

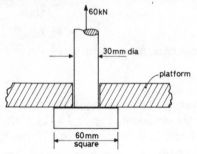

Fig. 13.13

5　In a block and tackle arrangement 6 ropes (each 20 mm diameter) support a mass of 100 kg. Calculate the stress in each rope.

6　A uniform beam of mass 200 kg and 2 m in length rests on two vertical pillars 90 mm × 90 mm square section (Figure 13.14). Calculate the compressive stress in each pillar.

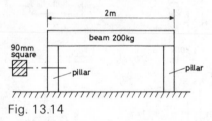

Fig. 13.14

7　A tie-bar 24 mm diameter is designed to a tensile force of 17 kN. Owing to corrosion the diameter of the bar is reduced to 22 mm. Calculate the percentage increase in stress for the same force.

8　A metal bar, original length 2·5 m, extends by 0·5 mm. Calculate the tensile strain.

9　A compressive load causes a support to shorten by 1·5 mm. Calculate the original length of the support if the compressive strain is $0·4 \times 10^{-3}$.

10　A copper bar 5 m long and 40 mm × 30 mm rectangular cross-section is stretched 2·25 mm by a tensile load of 60 kN. Calculate:

a) The tensile stress in the bar.
b) The tensile strain.
c) Young's Modulus of Elasticity for the material.

11 A vertical suspension cable of a bridge is made of steel wires wrapped into a bundle 60 mm in diameter. Calculate the extension to be expected in 15 m length when carrying a load of 200 kN. Take Young's Modulus as 200 GN/m^2.

12 Sections through three steel girders are shown in Figure 13.15. Each girder rests vertically on a concrete foundation and supports a load of 120 kN. Calculate by how much each girder shortens. Take E = 200 GN/m^2. Compare results.

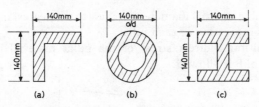

all thicknesses 20mm   each length 3m

Fig. 13.15

13 Four aluminium tubular rods (each 30 mm outside diameter) support a load of 32 kN. If the load is equally distributed and the direct stress in each rod is 20·4 MN/m^2, calculate:
a) The inside diameter of each rod.
b) The original length of each rod if the extension is 1 mm.
   Young's Modulus: 70 GN/m^2

14 If the mild steel spindle shown (Figure 13.16) is subjected to a compressive load of 120 kN, determine:
a) The compressive stress in each portion (1) and (2).
b) The total compression.
   E = 200 GN/m^2

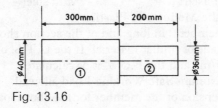

Fig. 13.16

15   If the ultimate tensile stress of mild steel is taken as 430 MN/m²
     determine for the same material:
     a) The maximum allowable load on a 55 mm diameter rod, allow-
         ing a factor of safety of 6.
     b) The extension under the above conditions on a 3·5 metre
         length of rod if Young's Modulus for mild steel is 200 GN/m².

16   Two alloy tubular pillars, each 60 mm outside diameter, have a
     dual purpose of supporting part of an overhanging deckhouse on
     board a ship, and to assist in drainage (Figure 13.17). Both pillars
     support equally a total load of 100 kN. If the S.W.S. of the
     material is 31·8 MN/m² and Young's Modulus 106 GN/m²,
     calculate:
     a) A suitable internal diameter of pillar (to the nearest millimetre).
     b) The compression of each pillar.
     c) The ultimate direct stress of the pillar material if a safety
         factor of 8 is allowed.

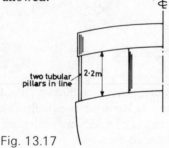

Fig. 13.17

17   *a*) The following information was taken from a framed structure
         calculation:

Length of member:	3 m
Extension of member:	0·6 mm
Diameter of member:	50 mm
Force in member:	78·5 kN
Nature of the force in member:	Tie
Material used:	Mild steel

         Calculate the Modulus of Elasticity.
     b) Another member 2 m long and of the section shown in Figure
         13.18, is made of similar material. If the U.T.S. of the material
         is 450 MN/m² and the F.O.S. is 5, calculate:
         i) the maximum Safe Working Load allowed.
         ii) the extension of the member for the above load.

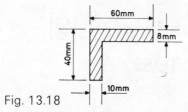

Fig. 13.18

18 The following readings were taken during a tensile test on a mild steel specimen. Draw the Load/Extension graph and determine from the graph the Modulus of Elasticity for the material.
Gauge length of specimen: 50 mm
Original diameter: 11·3 mm

load (kN)	8	12	16	20	28	34	40
extension (mm)	0·02	0·03	0·04	0·05	0·08	0·11	0·15

19 The following particulars were noted from a tensile test on mild steel.

Test diameter: 5·64 mm   Gauge length: 26 mm
Maximum load: 11 kN   Yield load: 5 kN
Final gauge length: 30·5 mm   Diameter at fracture: 3·8 mm
Determine:

a) The ultimate tensile stress.
b) The yield stress.
c) The percentage elongation.
d) The percentage reduction in area.

20 The following values were obtained from a tensile test on a metal bar test piece of 20 mm diameter and length 650 mm.

load (kN)	30	50	70	90	110	114	116	125
extension (mm)	0·3	0·5	0·7	0·9	1·1	1·3	1·6	2

130	140	144	140
2·3	3	3·6	5

a) Plot the Load/Extension diagram and on it mark the proportional limit and yield point.
b) Determine Young's Modulus for the material.
c) Determine the ultimate tensile stress if the maximum load is 145 kN.

14
Strength of Materials (2)
Shear Stress

Shear A material is said to be under shearing action when a load tends to make one layer of the material slide over an adjoining layer (Figure 14.1).

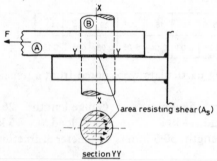

A load (F) applied to plate A tends to shear circular rod B. The combined effect of the internal resistance of rod B acts at right angles to its axis XX.

Fig. 14.1

Shear Stress τ is the internal reaction set up within a material due to it being subjected to a shearing action.

$$\text{Shear Stress} = \frac{\text{Load (N)}}{\text{Area resisting shear (m}^2)}$$

$$\tau = \frac{F}{A_S}$$

F = Applied (shearing) load
A_S = Area resisting shear
τ = Shear stress

(τ is the Greek letter 'tau'.)

A test under shear conditions can be carried out on a specimen of metal and a Shear Load/Deformation graph drawn, from which the Maximum Shear Load and Ultimate Shear Stress can be obtained. Shear deformation is proportional to Shear Load up to the proportional limit.

$$\text{Ultimate Shear Stress} = \frac{\text{Maximum (shear) Load}}{\text{Area resisting shear}} \text{ N/m}^2$$

$$\text{U.S.S.} = \frac{\text{M.L.}}{\text{A}_S}$$

$$\text{Safe Working (shear) Stress} = \frac{\text{Safe Working (shear) Load}}{\text{Area resisting shear}} \text{ N/m}^2$$

$$\text{S.W.S.} = \frac{\text{S.W.L.}}{\text{A}_S}$$

$$\text{Factor of Safety} = \frac{\text{Maximum (shear) Load}}{\text{Safe Working (shear) Load}}$$

$$\text{F.O.S.} = \frac{\text{M.L.}}{\text{S.W.L.}}$$

$$\text{Factor of Safety} = \frac{\text{Ultimate Shear Stress}}{\text{Safe Working (shear) Stress}}$$

$$\text{F.O.S.} = \frac{\text{U.S.S.}}{\text{S.W.S.}}$$

Single and Double Shear   Due to a shearing action materials tend to shear across one or two parallel planes. If the material tends to shear across one plane 'A.A' as shown in Figure 14.2a it is said to be in *Single Shear*. If the material tends to shear across two planes 'B.B' and 'C.C' as shown in Figure 14.2b it is said to be in *Double Shear*

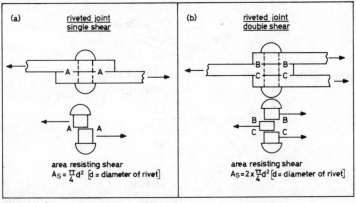

Fig. 14.2

Shear in Punching   The punch exerts a shearing force on the material and failure by shearing takes place along the perimeter of the punched hole; therefore the area of material resisting shear is the perimeter of the punched hole × thickness of the material (Figure 14.3).

Area Resisting Shear = Perimeter of punch × thickness of metal

$$A_S = P_r \times t$$

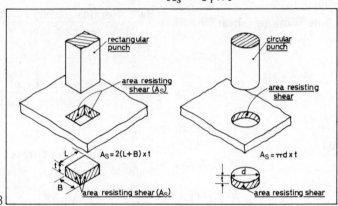

Fig. 14.3

Worked Examples

1 A piece of metal 100 mm wide × 5 mm thick is sheared across its width XX. If the force exerted by the shears is 70 kN calculate the failing stress of the metal.

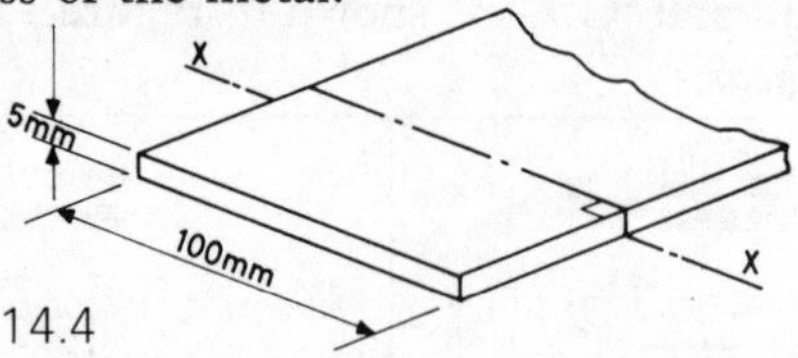

Fig. 14.4

Area resisting shear

$$A_S = 100 \times 5 \text{ mm}^2$$
$$= 500 \text{ mm}^2$$

$$\text{Shear stress} = \frac{\text{Load}}{\text{Area}}$$

$$\tau = \frac{F}{A_S}$$

$$\frac{70\,000}{500}$$

$$= 140 \text{ N/mm}^2$$
$$= 140 \text{ MN/m}^2$$

The failing stress of the metal is 140 MN/m²

2   In a steelworks, when cutting 20 mm diameter rods, the average failing stress in each section was $60 \, MN/m^2$. If 6 rods were cut in one operation calculate the total force exerted by the shears.

The area resisting shear is the cross-section of each rod.
Cross-sectional area of one rod

$$A = \frac{\pi}{4} d^2$$

$$= \frac{\pi}{4} \times 20 \times 20 \, mm^2$$

$$= 100\pi \, mm^2$$

$$= 314 \, mm^2$$

Total cross-sectional area of the rods

$$A_S = 6 \times 314 \, mm^2$$

$$= 1\,884 \, mm$$

$$= 1\,884 \times 10^{-6} \, m^2$$

$$\tau = \frac{F}{A_S}$$

$$F = \tau \times A_S$$

$$= 60 \times 10^6 \times 1\,884 \times 10^{-6} \, N$$

$$= 113\,000 \, N$$

$$= 113 \, kN$$

Force exerted by the shears $= 113 \, kN$

3   Calculate the maximum thickness of plate which can be sheared on a guillotine if the U.S.S. of the plate is $100 \, N/mm^2$ and the maximum force exerted by the blade of the guillotine is 200 kN. The width of the plate is 200 mm.

Let 't' be the thickness of the plate.

Area resisting shear $A_S = (200 \times t) \, mm^2$

Ultimate shear stress U.S.S. $= \dfrac{M.L.}{A_S}$

$$A_S = \frac{M.L.}{U.S.S.}$$

$$200 \times t = \frac{200 \times 10^3}{100} \, mm^2$$

$$t = \frac{200 \times 10^3}{100 \times 200} \, \text{mm}$$

$$t = 10 \, \text{mm}$$

Maximum thickness of plate which can be sheared = 10 mm.

4   A mild steel tie-bar 70 mm wide × 5 mm thick is connected to an angle bar by two 10 mm diameter rivets. If the maximum safe working load acting on the riveted connection is 15 kN calculate:

a) The maximum safe working (shear) stress in the rivet material.
b) The U.S.S. of the rivet material if the F.O.S. is 4.
c) The greatest tensile stress in the tie-bar.
d) The U.T.S. of the tie-bar for a factor of safety of 8.

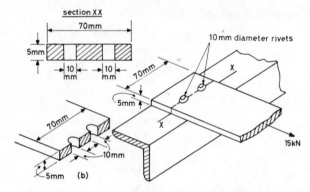

Fig. 14.5

a) Each rivet is in single shear.

Total area resisting shear = Number of rivets × cross-sectional area of one rivet.

$$A_S = 2 \times \frac{\pi}{4} \times d^2$$

$$= 2 \times \frac{\pi}{4} \times 10 \times 10 \, \text{mm}^2$$

$$= 50\pi \, \text{mm}^2$$

$$= 157 \, \text{mm}^2$$

$$\text{Shear stress } \tau = \frac{F}{A_S}$$

$$= \frac{15\,000}{157} \, \text{N/mm}^2$$

$$= 95 \cdot 5 \, \text{N/mm}^2$$
$$= 95 \cdot 5 \, \text{MN/m}^2$$

Maximum safe working shear stress in the rivet material = 95·5 MN/m²

b)
$$\text{F.O.S.} = \frac{\text{U.S.S.}}{\text{S.W.S.}}$$
$$\text{U.S.S.} = \text{F.O.S.} \times \text{S.W.S.}$$
$$= 4 \times 95 \cdot 5 \, \text{MN/m}^2$$
$$= 382 \, \text{MN/m}^2$$

The ultimate shear stress of the rivet material = 382 MN/m².

c) To determine the greatest tensile stress in the tie-bar consider
a section XX through the line of rivets (Figure 14.5b).
The greatest tensile stress occurs across the section of
minimum cross-sectional area.
Minimum cross-sectional area

$$\text{A}_{\text{min}} = (70 \times 5) \, \text{mm}^2 - 2(10 \times 5) \, \text{mm}^2$$
$$= 350 \, \text{mm}^2 - 100 \, \text{mm}^2$$
$$= 250 \, \text{mm}^2$$

$$\text{Greatest } \sigma = \frac{\text{F}}{\text{A}_{\text{min}}}$$
$$= \frac{15\,000}{250} \, \text{N/mm}^2$$
$$= 60 \, \text{N/mm}^2$$
$$= 60 \, \text{MN/m}^2$$

The greatest tensile stress in the tie-bar = 60 MN/m².

d)
$$\text{F.O.S.} = \frac{\text{U.T.S.}}{\text{S.W.S.}}$$
$$\text{U.T.S.} = \text{F.O.S.} \times \text{S.W.S.}$$
$$= 8 \times 60 \, \text{MN/m}^2$$
$$= 480 \, \text{MN/m}^2$$

The ultimate tensile stress of the tie-bar = 480 MN/m².

5   The figure shows the plan and section of a riveted connection.
Two plates A and B are connected to an angle beam by means of
a strap plate C. The rivets are 10 mm diameter and the failing
shear stress of the rivet material is 385 MN/m². Calculate:
a) The load F required to fracture the rivets.
b) The greatest tensile stress in plate B.

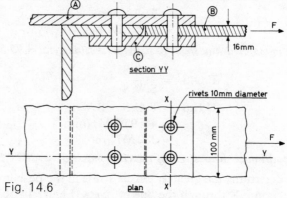

Fig. 14.6   <u>plan</u>

a) The rivets are in double shear.

Total number of rivets to shear = 2

Total area resisting shear

$$A_S = 2 \times 2 \times \frac{\pi}{4} d^2$$

$$= 2 \times 2 \times \frac{\pi}{4} \times 10 \times 10 \, \text{mm}^2$$

$$= 100\pi \, \text{mm}^2$$

$$= 314 \, \text{mm}^2$$

Ultimate shear stress $\tau = \dfrac{F \text{ (maximum load)}}{A_S}$

$$F = \tau \times A_S$$
$$= 385 \times 10^6 \times 314 \times 10^{-6} \, \text{N}$$
$$= 121\,000 \, \text{N}$$
$$= 121 \, \text{kN}$$

The load required to fracture the rivets = 121 kN.

b) Consider the section of the middle plate B through the line of rivets at XX.

Minimum cross-sectional area:

$$A_{min} = (100 \times 16) \, \text{mm}^2 - 2(10 \times 16) \, \text{mm}^2$$
$$= 1\,600 \, \text{mm}^2 - 320 \, \text{mm}^2$$
$$= 1\,280 \, \text{mm}^2$$

Greatest $\sigma = \dfrac{F}{A_{min}}$

$$= \frac{121\,000}{1\,280} \, \text{N/mm}^2$$

$$= 94.5 \, \text{N/mm}^2$$
$$= 94.5 \, \text{MN/m}^2$$

The greatest tensile stress in plate B $= 94.5 \, MN/m^2$

6   A triangular shaped plate carries two parallel loads, each of 20 kN as shown in Figure 14.7. The plate is connected to two flat metal bars by means of two 12 mm diameter rivets. Neglecting the mass of the plate calculate the shear stress in the rivet material.

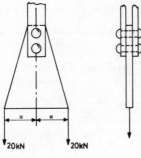

20kN   20kN

Fig. 14.7

Area resisting shear A_S = 2 rivets in double shear

$$= 2 \times 2 \times \frac{\pi}{4} \text{d}^2 \, \text{mm}^2$$

$$= \pi \times 12 \times 12 \, \text{mm}^2$$
$$= 144\pi \, \text{mm}^2$$
$$= 452 \, \text{mm}^2$$

Resultant load acting on the rivets $= 2 \times 20 \, \text{kN}$
$$= 40 \, \text{kN}$$

$$\tau = \frac{F}{A_S}$$

$$= \frac{40\,000}{452} \text{N/mm}^2$$

$$= 88.5 \, \text{MN/m}^2$$

Shear stress in the rivet material $= 88.5 \, MN/m^2$

7   A knuckle joint forming part of a tie-bar is shown. If an axial pull of 50 kN is applied, the U.S.S. for the material is 385 MN/m² and the F.O.S. is 7, calculate:

a) The tensile stress in rods (1) and (2).

b) The diameter of the pin required for this load.

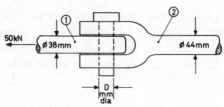

Fig. 14.8

a) To find the tensile stress in rod (1)

Cross-sectional area: $A_X = \dfrac{\pi}{4}d^2$

$$= \frac{\pi}{4} \times 38 \times 38 \, \text{mm}^2$$

$$= 1\,134 \, \text{mm}^2$$

Tensile stress $= \dfrac{F}{A_X}$

$$= \frac{50\,000}{1\,134} \, \text{N/mm}^2$$

$$= 44\cdot1 \, \text{N/mm}^2$$
$$= 44\cdot1 \, \text{MN/m}^2$$

The tensile stress in rod (1) = 44·1 MN/m².

To find the tensile stress in rod (2)

Cross-sectional area: $A_X = \dfrac{\pi}{4}d^2$

$$= \frac{\pi}{4} \times 44^2 \, \text{mm}^2$$

$$= 1\,520 \, \text{mm}^2$$

Tensile Stress $\sigma = \dfrac{F}{A_X}$

$$= \frac{50\,000}{1\,520}\,\mathrm{N/mm^2}$$

$$= 32 \cdot 9\,\mathrm{N/mm^2}$$

$$= 32 \cdot 9\,\mathrm{MN/m^2}$$

The tensile stress in rod (2) = 32·9 MN/m².

b) To find the diameter 'D' of the pin.
   The pin is in double shear.
   Total area resisting shear

$$A_S = 2\left(\frac{\pi}{4}D^2\right)\,\mathrm{mm^2}$$

$$= \frac{\pi}{2}D^2\,\mathrm{mm^2}$$

$$= 1 \cdot 57\,D^2\,\mathrm{mm^2}$$

Factor of safety

$$\mathrm{F.O.S.} = \frac{\mathrm{U.S.S.}}{\mathrm{S.W.S.}}$$

$$\mathrm{S.W.S.} = \frac{\mathrm{U.S.S.}}{\mathrm{F.O.S.}}$$

$$= \frac{385}{7}\,\mathrm{MN/m^2}$$

$$= 55\,\mathrm{MN/m^2}$$

And $\mathrm{S.W.S.} = \dfrac{F}{A_S}$

$$55\,\mathrm{N/mm^2} = \frac{50\,000}{1 \cdot 57\,D^2}$$

$$D^2 = \frac{50\,000}{1 \cdot 57 \times 55}\,\mathrm{mm^2}$$

$$D^2 = 579\,\mathrm{mm^2}$$

$$D = 24 \cdot 1\,\mathrm{mm}$$

The diameter of the pin required = 24·1 mm.

8 A 20 mm diameter hole is punched in a sheet of metal 3 mm thick.
   If the U.S.S. of the metal is 152 MN/m² calculate:
   a) The least force that has to be applied to the punch in order to
      cut the metal.
   b) The direct stress in the punch for the above force.

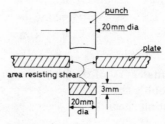

Fig. 14.9

a) Area resisting shear:

$$A_S = \text{Perimeter of the punch} \times \text{thickness of metal}$$
$$= \pi d \times t$$
$$= 20\pi \times 3 \text{ mm}^2$$
$$= 188\cdot4 \text{ mm}^2$$
$$= 188\cdot4 \times 10^{-6} \text{ m}^2$$

Shear stress:

$$\tau = \frac{F}{A_S}$$
$$F = \tau \times A_S$$
$$= 152 \times 10^6 \times 188\cdot4 \times 10^{-6} \text{ N}$$
$$= 28\,650 \text{ N}$$
$$= 28\cdot65 \text{ kN}$$

The force applied to the punch = 28·65 kN.

b) To find the direct stress in the punch.

Cross-sectional area of the punch $A_X = \dfrac{\pi}{4}d^2$

$$= \frac{\pi}{4} \times 20 \times 20 \text{ mm}^2$$
$$= 100\pi \text{ mm}^2$$
$$= 314 \text{ mm}^2$$

Direct stress in the punch $= \dfrac{F}{A_X}$

$$= \frac{28\cdot65 \times 10^3}{314} \text{ N/mm}^2$$

$$= 91 \cdot 2 \, \text{N/mm}^2$$
$$= 91 \cdot 2 \, \text{MN/m}^2$$

The direct (compressive) stress in the punch = 91·2 MN/m².

9   A press stamps out metal washers 20 mm outside diameter, 10 mm
inside diameter and 1·9 mm thick. If the U.S.S. of the material is
140 MN/m² and the punching force exerted is 150 kN, calculate:
a) The maximum thickness of sheet that the punch can penetrate.
b) The number of washers that can be stamped out in one blow of
the punch.

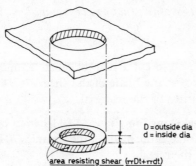

D = outside dia
d = inside dia

area resisting shear (πDt+πdt)

Fig. 14.10

a) Let 'D' be the outside diameter of the washer. Let 'd' be the
inside diameter of the washer.
Area resisting shear (Figure 14.10):

$$A_S = \pi Dt + \pi dt \qquad \qquad t = \text{thickness of plate in mm}$$
$$= 20\pi t \, \text{mm}^2 + 10\pi t \, \text{mm}^2$$
$$= 30\pi t \, \text{mm}^2$$
$$= 94 \cdot 2t \, \text{mm}^2$$

$$\text{U.S.S.} = \frac{\text{M.L.}}{A_S}$$

$$A_S = \frac{\text{M.L.}}{\text{U.S.S.}}$$

$$94 \cdot 2t = \frac{150\,000}{140}$$

$$t = \frac{150\,000}{140 \times 94 \cdot 2}$$

$$= 11 \cdot 4$$

The maximum thickness of metal that the punch will penetrate =
11·4 mm.

b) Number of washers that can be stamped out of sheets 1·9 mm
thick in one blow of the punch

$$= \frac{\text{Maximum penetration}}{\text{Thickness of washer}}$$

$$= \frac{11 \cdot 4 \, \text{mm}}{1 \cdot 9 \, \text{mm}}$$

$$= 6$$

The number of washers that can be stamped out in one blow = 6.

Examples

1   A guillotine is used to cut a strip of metal 30 mm wide by 3 mm
thick. If the force at the blade of the shears is 9 kN calculate the
ultimate shear stress of the material.

2   The bar of metal shown in Figure 14.11 has to be sheared along
line XX. If the force at the blade of the shears is 70 kN calculate the
failing stress of the metal.

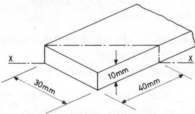

Fig. 14.11

3   In a steel mill, rods of various cross-sections are sheared off to
lengths (Figure 14.12). If the rods are made of the same material of
U.S.S. $140 \, \text{N/mm}^2$, calculate in each case the force required to
shear each rod.

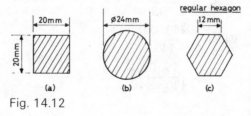

Fig. 14.12

4 A force of 120 kN is required to shear a number of rods, each 22·5 mm diameter. If the failing stress in each section is 60 MN/m² calculate the number of rods which can be sheared in one operation of the shears.

5 Figure 14.13:
   a) A single riveted connection is subjected to a load of 10 kN. Calculate the shear stress in the 20 mm diameter rivet.
   b) Calculate the shear stress in the rivet material for the connection shown. The rivets are 14 mm diameter and are subjected to a load of 20 kN.

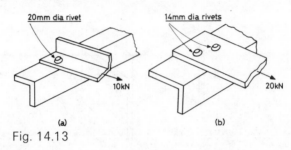

(a)   (b)

Fig. 14.13

6 The double lap riveted joint shown in Figure 14.14 is subjected to a load of 80 kN. The shear stress in each rivet is not to exceed 60 MN/m². Calculate the diameter of the rivets for maximum allowable shear stress.

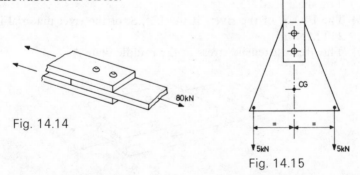

Fig. 14.14

Fig. 14.15

7 A triangular plate (mass 51 kg) of uniform thickness carries two parallel loads, each of 5 kN as shown in Figure 14.15. The plate is connected to a vertical member by two rivets in single shear. If the shear stress in the rivet material is not to exceed 83 MN/m² calculate the diameter of the rivets.

8   In the design of a double lap riveted joint the U.S.S. of the rivet material is $240 \, MN/m^2$. The F.O.S. allowed is 4 and the load acting on the connection is 24 kN. Calculate the number of 8 mm diameter rivets required.

9   A bin containing waste metal is connected to two leg supports A and B by eight 14 mm diameter rivets. Assuming that the total load of the bin and its contents is shared equally by the supports, calculate for a total mass of 8 tonne, the shear stress in the rivet material connecting leg A (Figure 14.16).

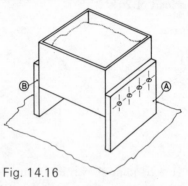

Fig. 14.16

10   A double lap joint is secured by five 10 mm diameter rivets. The shear stress in each rivet is $40 \, MN/m^2$ (Figure 14.17). Calculate:
   a) The shear load F.
   b) The F.O.S. of the rivets if the U.S.S. of the rivet material is $240 \, MN/m^2$.
   c) The greatest tensile stress in the middle plate A.

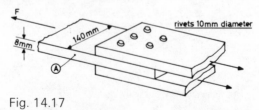

Fig. 14.17

11   A plate (mass 510 kg) is held in position as shown in Figure 14.18. The shackle pin is 10 mm diameter and the rope supporting the load is 12 mm diameter. Calculate:
   a) The shear stress in the shackle pin.
   b) The tensile stress in the rope.

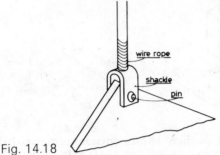

Fig. 14.18

2   The framed structure shown (Figure 14.19) carries a load of 20 kN.
Determine:
   a) The force in member A.
   b) The shear stress in the pin at connection B (inset diagram).
   c) The greatest direct stress in plate connection C.

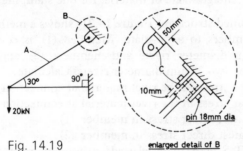

Fig. 14.19   enlarged detail of B

3   A punching machine can exert a force of 77 kN. What is the
greatest diameter hole which can be punched in material 10 mm
thick if the U.S.S. of the material is 100 MN/m^2?

4   If the ultimate shear stress of a metal is taken as 100 MN/m^2
find the force necessary to punch a hole:
   a) 20 mm diameter in a plate 5 mm thick.
   b) A rectangular hole 20 mm × 10 mm in a plate 5 mm thick.
       Find also the compressive stress in the punch in each case.

15 A square hole (20 mm side) is punched in a square sheet of metal
100 mm × 100 mm × 4 mm thick (Figure 14.20). If the U.S.S. of
the material is 140 MN/m^2 calculate:
   a) The force F_1 exerted by the punch.
   b) The compressive stress in the punch.
   c) The S.W.L. F_2 if the S.W.S. in the metal is 80 MN/m^2.

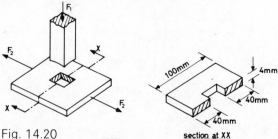

Fig. 14.20   section at XX

16 A press stamps out washers 18 mm outside diameter, 10 mm inside diameter and 1·7 mm thick. If the U.S.S. of the material is 200 N/mm² and the punching force is 150 kN, calculate:
a) The maximum thickness of material that can be punched.
b) The number of washers that can be stamped out in one operation of the punch.
c) The percentage waste of material for one stamping operation.

17 Part of a framed structure (Figure 14.21) shows a method of connecting members to a tie-plate. Member ① is connected by three 20 mm diameter rivets and members ② and ③ each connected by two 20 mm diameter rivets. Calculate:
a) The shear stress in the rivet material at connection A.
b) The shear stress in the rivet material at connection B.
c) The greatest direct stress in member ① .
d) The greatest direct stress in member ② .
If the U.T.S. of the material used for the members is 455 MN/m²
and the U.S.S. of the rivet material is 348 MN/m², determine:
e) The F.O.S. for connection A.
f) The F.O.S. for the material in member ② .

18 Two shafts are joined together by a flanged coupling. The six 12 mm diameter bolts securing the coupling share equally the total tangential load and are spaced equidistant on a pitch circle of 200 mm diameter (Figure 14.22). If the torque transmitted is 5 000 Nm calculate the shear stress in each bolt.

19 In the design of a flanged coupling, similar to that in Fig. 14.22, the bolts are to be 16 mm diameter and are to be equally spaced on a pitch circle of 280 mm diameter. The torque transmitted is to be 5·4 kNm, the ultimate shear stress of the bolt material is 384 MN/m², and there is to be a factor of safety of 8.
Calculate the number of bolts required.

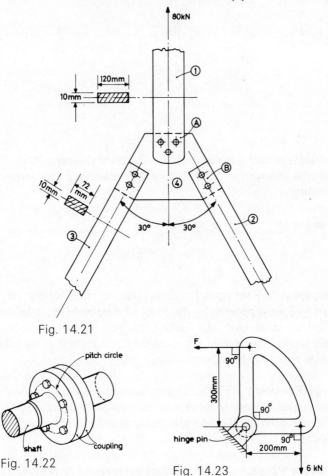

Fig. 14.21

Fig. 14.22

Fig. 14.23

20 A quadrant bracket is supported in the position shown by a
    force F (Figure 14.23). Neglect the mass of the quadrant:
    a) Calculate the force F to maintain equilibrium.
    b) If the hinge pin is 8 mm diameter and in double shear calculate
       the shear stress in the pin.
    c) If the U.S.S. of the pin material is 360 MN/m² and a F.O.S.
       of 5 allowed, determine whether the pin would be within the
       safe working limit.
    d) By using a F.O.S. of 4 determine the diameter of a pin made of
       similar material.

15
Fluids

Both liquids and gases are included under the term *fluids*. The particular property of fluids is that they offer very little resistance to change of shape.

Pressure (P) is force on unit area, *i.e.*

$$P = \frac{F}{A} (N/m^2)$$

Atmospheric Pressure (P_{AT}) is due to the weight of the column of air supported by the particular point being considered. Its value varies over the earth's surface (climatic conditions, height above sea level) and a standard value is used. This value is $101\,000\,N/m^2$ or $0 \cdot 1\,MN/m^2$.

Gauge Pressure The reading on a pressure gauge (*e.g.* bourdon gauge, manometer) is known as the *gauge pressure* and is the difference between the pressure being measured and atmospheric pressure.

Absolute Pressure is the actual pressure of the fluid. Thus: Absolute pressure = Gauge pressure + atmospheric pressure.

Density (ρ) The density of a substance is the mass of unit volume (kg/m^3). ρ is the Greek letter 'rho'.

Specific Weight (w) is the weight of unit volume, *i.e.*

$$w = \rho g$$

Specific Gravity (s) is the ratio of the mass of a substance to the mass of an equal volume of some standard substance. The standard substance used is normally water.

$$s_{fluid} = \frac{m_{fluid}}{m_{water}} = \frac{\rho_{fluid}}{\rho_{water}}$$

Axioms

1 The pressure at a point is the same in all directions.
2 The pressure in a liquid increases uniformly with depth.
3a) The free surface of a liquid at rest is a horizontal plane.
 b) The common surface of two liquids of different densities is a horizontal plane.
4 In an enclosed vessel (*e.g.* a gas cylinder) the pressure is uniform throughout and acts at right angles to the walls of the vessel (Figure 15.1).

Fig. 15.1

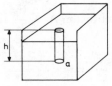

Fig. 15.2

Pressure at a Depth   is due to a 'head' of liquid. Consider an area a(m^2) at a depth h(m) in a liquid of density ρ (Figure 15.2).

The volume of liquid supported on 'a' = ha (m^3)
Mass of this liquid $= \rho$ha (kg)
i.e. The force on the area $= \rho$hag (N)

Thus the pressure on the area $= \dfrac{\rho hag}{a}$ (N/m^2)

$= \rho$gh (N/m^2)

∴ In any liquid the pressure due to a head h of the liquid,

$$P = \rho gh \quad or \quad = wh$$

NOTE: This is the pressure due to the liquid only. The absolute pressure will be found by adding on the atmospheric pressure, since atmospheric pressure acts on the surface of the liquid.

Principle of Archimedes   If a body is totally or partially immersed in a fluid it receives an upthrust which is equal to the weight of fluid displaced.

Hydraulic (Bramah) Press

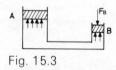

Fig. 15.3

When two connected cylinders are filled with a liquid and have close fitting pistons (at A and B Figure 15.3), then by exerting a force on one piston we can create a force at the other piston, *i.e.* overcome a resistance. Let 'a' be the area of piston A, and 'b' be the area of piston B. Let a force F_B be exerted on piston B.

$$\text{Pressure at B} = \frac{F_B}{b}$$

But since the pressure is uniform throughout the liquid, the pressure at piston $A = \dfrac{F_B}{b}$ and thus the force exerted at piston A

$$F_A = \text{Pressure} \times \text{area}$$

$$= \frac{F_B}{b} \times a$$

i.e. $\quad F_A = F_B \times \dfrac{a}{b}$

By making 'a' large and/or 'b' small we can exert a much greater force at A than the initial force at B. This is the basic principle of all hydraulic presses, jacks and other lifting equipment.

Boyle's Law   If a fixed mass of any gas is compressed at constant temperature, then the volume is inversely proportional to the absolute pressure, *i.e.*

$$V \propto \frac{1}{P}$$

$$V = C \times \frac{1}{P} \quad \text{(where C is a constant}$$
$$or \quad PV = C \qquad \text{for the given conditions)}$$

Pressure Gauges
Simple Barometer (Figure 15.4)
The simple barometer is used for measuring atmospheric pressure. The liquid used is mercury, because of its high specific gravity,

and the height h of the column is a direct measure of the pressure.
Thus standard atmospheric pressure is:

$$0 \cdot 1 \text{ MN/m}^2$$
$$or \quad 760 \text{ mm of mercury (Hg)}$$

Aneroid Barometer
The simple mercury barometer suffers the disadvantage that because of its size and the open liquid it is not readily portable. It is often replaced by an aneroid barometer on which the pressures are read directly from a dial.

Simple U Tube Manometer (Figure 15.5)
This gauge consists of a U tube containing a fluid of higher specific gravity than that to be measured. Let ρ_1 be the density of the manometer fluid, and ρ_2 be the density of the fluid to be measured. Take $P_4 (= P_{AT})$ as datum. Since pressure at $X = P_1$ and $P_2 = P_3$ then relative to P_{AT}, $P_3 = \rho_1 gh = P_2$

$$\text{and} \quad P_1 = \rho_1 gh - \rho_2 gx = P_x$$
$$i.e. \quad P_x = \rho_1 gh - \rho_2 gx \text{ (gauge)}$$

NOTE: If the unknown pressure is that of a gas then ρ_2 will be very small and the second term may be neglected.

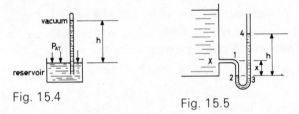

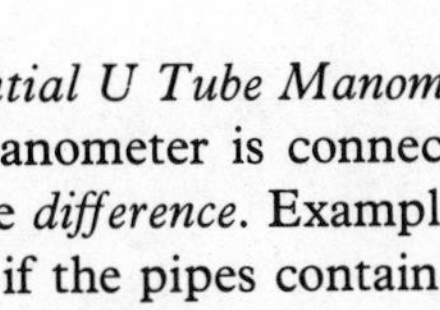

Fig. 15.4

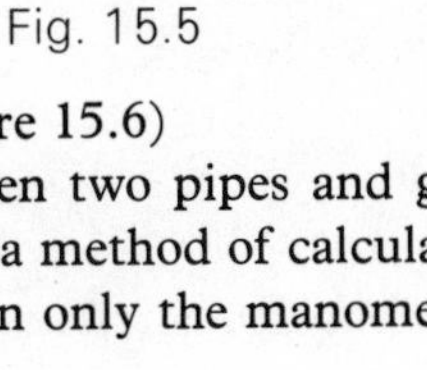

Fig. 15.5

Differential U Tube Manometer (Figure 15.6)
This manometer is connected between two pipes and gives the pressure *difference*. Example 5 shows a method of calculation. As before, if the pipes contain a gas, then only the manometer fluid need be considered.

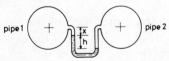

Fig. 15.6

Bourdon Pressure Gauge
The bourdon gauge is similar in construction to the aneroid barometer, and is almost universally used. It has the advantages that it is small, convenient to use and can be used up to very high pressures.

Worked Examples

1 A simple mercury barometer is used to obtain the atmospheric pressure.

   a) The length of the mercury column is 770 mm. Calculate from first principles the corresponding pressure in N/m^2 (density of Hg is $13\cdot6 \times 10^3$ kg/m^3).

   Let the cross section of the tube be a m^2.
   The volume of mercury $= 0\cdot77 \times a\,m^3$
   And the mass of mercury $= 0\cdot77 \times a \times 13\cdot6 \times 10^3$ kg
   $\therefore$ Weight of mercury $= 0\cdot77 \times a \times 13\cdot6 \times 10^3 \times g$ N

   $$P = \frac{F}{A} = \frac{0\cdot77 \times a \times 13\cdot6 \times 10^3 \times g}{a} \, N/m^2$$

   $$= 0\cdot77 \times 13\cdot6 \times 10^3 \times 9\cdot81 \, N/m^2$$

   $$\therefore \quad P_{AT} = 102\cdot8 \times 10^3 \, N/m^2$$

   i.e. Atmospheric pressure $= 0\cdot1028 \, MN/m^2$

   b) If the atmospheric pressure drops to $0\cdot102$ MN/m^2, calculate the new length of the mercury column.

   $$P_{AT} = 102 \times 10^3 \, N/m^2 \qquad P = \rho g h$$

   $$h = \frac{P_{AT}}{\rho g}$$

   $$= \frac{102 \times 10^3}{13\cdot6 \times 10^3 \times 9\cdot81}$$

   $$= 0\cdot765 \, m$$

   $\therefore$ New length of column is 765 mm.

2 A water tank 3 m square contains fresh water to a depth of $2\cdot5$ m. Determine:

   a) The pressure at the bottom of the tank due to the water.

   b) The actual pressure at the bottom of the tank.

   c) The resultant force on the bottom of the tank if it is supported clear of the ground.

a) Due to the water alone, $P = wh$

$$= \rho g h$$

$$= 1 \times 10^3 \times 9\cdot81 \times 2\cdot5 \ N/m^2$$

i.e. $P = 24\cdot55 \times 10^3 \ N/m^2$

b) The actual pressure at the bottom of the tank

$$= 24\cdot55 \times 10^3 + P_{AT} \ N/m^2$$

$$= 24\cdot55 \times 10^3 + 100 \times 10^3 \ N/m^2$$

$\therefore$ *Actual pressure* $= 124\cdot55 \times 10^3 \ N/m^2$

c) If the tank is clear of the ground there will be air pressure under the tank which will neutralise that above the water. The resultant force on the bottom will thus be due to the water alone.

Total force exerted $= P \times A$

$$= 24\cdot55 \times 10^3 \times (3 \times 3) \ newtons$$

$$= 221 \times 10^3 \ newtons$$

i.e. Resultant force on the tank $= 221 \ kN$

3   A tank holds 500 kg of an oil A of specific gravity 0·9. Determine the mass of an oil B, of specific gravity 0·8, that the tank will hold.

Density of water $= 1 \times 10^3 \ kg/m^3$

$\therefore$ Density of oil A $= 0\cdot9 \times 1 \times 10^3 \ kg/m^3$

Volume of the tank $= \dfrac{m}{\rho} = \dfrac{500}{0\cdot9 \times 10^3} \ m^3$

Density of oil B $= 0\cdot8 \times 10^3 \ kg/m^3$

$\therefore$ Mass of oil B $= V \times \rho$

$$= \frac{500}{0\cdot9 \times 10^3} \times 0\cdot8 \times 10^3 \ kg$$

$$= 500 \times \frac{0\cdot8}{0\cdot9} \ kg \ \left(i.e. \ \text{Mass} \times \frac{s_B}{s_A} \right)$$

i.e. Mass of oil B $= 445 \ kg$

4   (Figure 15.7) A simple mercury filled U tube manometer is used to measure the pressure of oil flowing through a pipe. The specific gravity of the oil is 0·8. Calculate the pressure, relative to the atmosphere of the oil at A:

a) in mm of mercury.

b) in N/m^2.

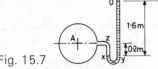

Fig. 15.7

a)   Pressure at $0 = 0$ (atmospheric pressure)

$$P_y = P_x = 1\,600 \text{ mm Hg}$$

$$P_z = P_x - h_{xz}$$

Now, $h_{xz} = 200$ mm of oil

$$= 200 \times \frac{s_{oil}}{s_{Hg}} \text{ mm Hg}$$

$$= 200 \times \frac{0 \cdot 8}{13 \cdot 6} \text{ mm Hg}$$

$$\therefore \quad P_z = 1\,600 - 200 \times \frac{0 \cdot 8}{13 \cdot 6} \text{ mm Hg}$$

i.e. $P_z = 1\,588$ mm Hg

$\therefore$ *Pressure head at A = 1 588 mm Hg*

b) $P_{oil} = \rho gh$

$$= 13 \cdot 6 \times 10^3 \times 9 \cdot 81 \times 1 \cdot 588 \text{ N/m}^2$$

i.e. $P_{oil} = 212 \times 10^3 \text{ N/m}^2$

5 (Figure 15.8) A differential U tube manometer containing carbon tetrachloride is used to measure the pressure difference between two water pipes. Determine this pressure difference in the situation shown (specific gravity of carbon tetrachloride 1·6).

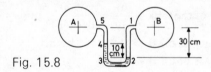

Fig. 15.8

Take pipe B as datum and use heads of water.

$$P_2 = P_1 + 30 \text{ cm H}_2\text{O}$$

$$P_2 = P_3$$

$$P_4 = P_3 - 10 \times \frac{1 \cdot 6}{1} \text{ cm H}_2\text{O}$$

$$= P_1 + 30 - 16 \text{ cm H}_2\text{O}$$

$$P_5 = P_4 - 20 \text{ cm H}_2\text{O}$$

$$= P_1 + 30 - 16 - 20 \text{ cm H}_2\text{O}$$

i.e. $P_5 = P_1 - 6 \text{ cm H}_2\text{O}$

i.e. The difference in pressure is a head of 6 cm of water, A being at the lower pressure.

$$P_{AB} = \rho gh$$

$$= 1 \times 10^3 \times 9{\cdot}81 \times 6 \times 10^{-2}$$
$$\therefore \; \textit{Pressure difference} = 590 \; N/m^2$$

6   A block of wood 1 m by 0·25 m has a mass of 12 kg. Find the depth
to which it will be immersed when placed in a tank containing:
a) salt water of density $1{\cdot}02 \times 10^3$ kg/m^3
b) oil of density $0{\cdot}8 \times 10^3$ kg/m^3

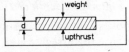

Fig. 15.9

a) Let 'd' be the depth of immersion.

The weight of the block is
$$12 \times 9{\cdot}81 \; N = 118 \; N$$
$$\therefore \; \text{Upthrust} = 118 \; N$$

i.e.

Volume of water × specific weight = 118 N
$$(1 \times 0{\cdot}25 \times d) \times (0{\cdot}8 \times 10^3 \times g) = 118$$
$$d = \frac{118}{1 \times 0{\cdot}25 \times 1{\cdot}02 \times 10^3 \times 9{\cdot}81} \; m$$
$$= 0{\cdot}047 \; m$$

i.e. The block will be immersed to a depth of 4·7 cm in salt water.

b) As above,

Volume of oil × specific weight = 118 N
$$(1 \times 0{\cdot}25 \times d) \times (0{\cdot}8 \times 10^3 \times g) = 118$$
$$d = \frac{118}{1 \times 0{\cdot}25 \times 0{\cdot}8 \times 10^3 \times 9{\cdot}81} \; m$$
$$= 0{\cdot}06 \; m$$

i.e. The block will be immersed to a depth of 6 cm in the oil.

7   A coal barge having a mass of 80 tonnes and a waterline area of
120 m^2 is loaded with 100 tonnes of coal in a fresh water dock.
Determine:
a) The depth of water required for the barge; assume the cross-
section to be uniform.
b) The change which will take place when the barge is towed into a
tidal estuary.

a) Let 'd' be the depth of immersion

$$\text{Upthrust} = \text{Weight of barge}$$

$$\therefore \text{ Volume} \times \text{specific weight} = (100 + 80) \times 10^3 \times 9 \cdot 81$$

$$(120 \times d) \times (1 \times 10^3 \times 9 \cdot 81) = 180 \times 10^3 \times 9 \cdot 81$$

$$d = \frac{180}{120 \times 1} \text{ m}$$

$$= 1 \cdot 5 \text{ m}$$

$\therefore$ *The barge requires a minimum depth of 1·5 m of water.*

b) When the barge moves into salt water it will rise slightly, the water being more dense than fresh water.

In this case,

$$(120 \times d) \times (1 \cdot 02 \times 10^3 \times 9 \cdot 81) = 180 \times 10^3 \times 9 \cdot 81$$

$$d = \frac{180}{120 \times 1 \cdot 02} \text{ m}$$

$$= 1 \cdot 47 \text{ m}$$

$\therefore$ *The new draught is 1·47 m; i.e. the barge has risen in the water by 30 mm.*

8 (Figure 15.10) A hydraulic jack is used to lift a beam section into position. The diameter of the load piston is 4 cm and of the effort piston 1 cm. Determine:

a) The pressure of the fluid if the load on the jack is 15 kN.

b) The effort F required at B.

c) The number of strokes of the effort piston required to raise the beam 25 mm, if the piston has a stroke of 50 mm.

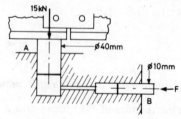

Fig. 15.10

a) Load (F) = 15 000 N   Area (A) = $\frac{\pi}{4}(4 \times 10^{-2})^2$

$$P = \frac{F}{A}   = 4\pi \times 10^{-4} \text{ m}^2$$

$$= \frac{15 \times 10^3}{4\pi \times 10^{-4}} \text{ N/m}^2$$

$$= \frac{150}{4\pi} \times 10^6 \text{ N/m}^2$$

i.e. Fluid pressure = 11·9 MN/m²

b) $\dfrac{F_A}{F_B} = \dfrac{\text{Area A}}{\text{Area B}}$ *or* Force = $P \times A$

$$= (11 \cdot 9 \times 10^6) \times \left(\frac{\pi}{4} \times 1^2 \times 10^{-4} \right)$$

$$\frac{15\,000}{F_B} = \frac{\dfrac{\pi}{4} \times 4^2}{\dfrac{\pi}{4} \times 1^2} \qquad F_B = 937 \text{ N}$$

$$\frac{15\,000}{F_B} = \frac{16}{1}$$

$$F_B = 937 \text{ N}$$

∴ *Force required at B is 937 N.*

c) If the load piston rises 25 mm, the volume of fluid required is

$$\frac{\pi}{4}(4 \times 10^{-2})^2 \times (25 \times 10^{-3}) \text{ m}^3 = 10\pi \times 10^{-6} \text{ m}^3$$

The amount of fluid/stroke transferred by the effort piston is

$$\frac{\pi}{4}(1 \times 10^{-2})^2 \times (50 \times 10^{-3}) \text{ m}^3 = \frac{5}{4}\pi \times 10^{-6} \text{ m}^3$$

The number of strokes is $\dfrac{10\pi \times 10^{-6}}{\dfrac{5}{4}\pi \times 10^{-6}}$

$$= 8$$

9 A vessel contains 3 m³ of a gas at atmospheric pressure. The gas is compressed until the pressure is 0·45 MN/m². Calculate the new volume of the gas.

$$P_1 = 0 \cdot 1 \text{ MN/m}^2 \qquad\qquad P_2 = 0 \cdot 45 \text{ MN/m}^2$$
$$V_1 = 3 \text{ m}^3 \qquad\qquad V_2 = ?$$
$$PV = C$$

or $P_1 V_1 = P_2 V_2 = P_3 V_3 = \; . \; . \; . \; . \; . \; = C$

Thus $0 \cdot 1 \times 3 = 0 \cdot 45 \times V_2$

$$V_2 = \frac{0 \cdot 1 \times 3}{0 \cdot 45} \text{ m}^3$$

$$= 0 \cdot 667 \text{ m}^3$$

i.e. The new volume of the gas is 0·667 m³.

10   The cylinder of an air compressor has a swept volume of $540 \times 10^{-6}\,m^3$ and a clearance volume of $80 \times 10^{-6}\,m^3$. If the reading on a pressure gauge is $20 \times 10^3\,N/m^2$ at the start of the stroke, determine the gauge reading at the end of the stroke.

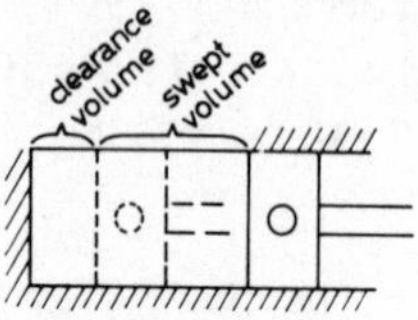

Fig. 15.11

$$\text{Absolute } P = \text{Gauge } P + P_{AT}$$
$$\therefore P_1 = (20 \times 10^3 + 100 \times 10^3)\,N/m^2$$
$$P_1 = 120 \times 10^3\,N/m^2$$
$$V_1 = \text{Swept volume} + \text{clearance volume}$$
$$= (540 \times 10^{-6} + 80 \times 10^{-6})\,m^3$$
$$\therefore V_1 = 620 \times 10^{-6}\,m^3$$
$$V_2 = \text{Clearance volume}$$
$$\therefore V_2 = 80 \times 10^{-6}\,m^3$$
$$P_1V_1 = P_2V_2$$
$$P_2 = \frac{P_1V_1}{V_2}$$
$$= \frac{120 \times 10^3 \times 620 \times 10^{-6}}{80 \times 10^{-6}}\,N/m^2$$
$$= 930 \times 10^3\,N/m^2 \text{ (absolute)}$$
$$\text{Absolute } P = \text{Gauge } P + P_{AT}$$
$$\text{Gauge } P = \text{Absolute } P - P_{AT}$$
$$= 930 \times 10^3\,N/m^2 - 100 \times 10^3\,N/m^2$$

i.e.   *The final reading on the pressure gauge is $830 \times 10^3\,N/m^2$.*

Examples

1 A mercury barometer reads 763 mm. Calculate:
   a) The atmospheric pressure in MN/m^2.
   b) The equivalent length of a water barometer.

2 A dam contains water to a depth of 20 m. Calculate:
   a) The pressure at the bottom due to the head of water.
   b) The actual pressure at the bottom.

3 (Figure 15.12) A tank contains water to a depth of 2·5 m. In the bottom is a circular valve V, 0·15 m in diameter, retained in position by the rod R. Determine the tension F in the rod required to keep the valve closed.

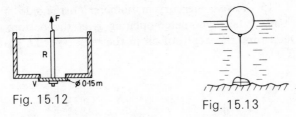

Fig. 15.12   Fig. 15.13

4 A punt 2·5 m long and 1·25 m wide has a mass of 204 kg. Find:
   a) The depth to which it will be immersed when placed in fresh water.
   b) The change which will occur when two men, each of mass 75·5 kg step into the punt.

5 (Figure 15.13) A marker float of diameter 0·2 m and mass 0·25 kg is moored in salt water with exactly half of the float below the surface. Calculate the tension in the mooring rope.

6 A fishing trawler has a displacement of 210 tonnes and a waterline area of 110 m². Determine the change which will take place when the trawler moves from a salt water basin into a fresh water canal.

7 A floating dock 60 m long and 12 m wide has a depth of 4 m. The mass of the dock is 1 500 t. Calculate the volume of salt water which must be pumped into the dock so that it will float with 0·5 m above the surface of the water.

8 (Figure 15.14) The pontoons of a bridge across a river each have a mass of 3 t and a cross-sectional area at the waterline of 12 m². determine the maximum amount that each pontoon will sink as a truck of mass 5 t is driven slowly across the bridge.

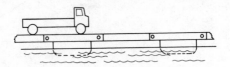

Fig. 15.14

9 A simple U tube mercury manometer is used to measure the pressure of a gas supply. If the difference in levels of the two limbs is 25 mm, determine the absolute pressure of the gas in N/m^2.

10 Figure 15.15 shows a mercury manometer on the side of an oil storage tank. The manometer opening is at the bottom of the tank. Determine the depth of oil in the tank (specific gravity of the oil is 0·9).

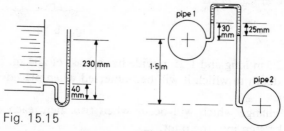

Fig. 15.15

Fig. 15.16

11 A differential manometer containing oil (s.g. 0·8) is used to measure the pressure difference between two gas pipes. The difference in the levels in the two limbs is 75 mm. Calculate the pressure difference.

12 (Figure 15.16) An inverted U tube manometer is used to measure the pressure difference between two water pipes as shown. Determine this difference, stating which is at the higher pressure. The density of the manometer liquid is $0·9 \times 10^3 \, kg/m^3$.

13 (Figure 15.17) A hydraulic ram is used to punch discs from a piece of sheet steel. The resistance of the steel is 2·5 kN. Determine:
   a) The pressure in the hydraulic fluid.
   b) The effort required at the master cylinder.

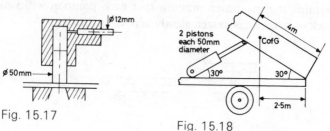

Fig. 15.17

Fig. 15.18

14 A hydraulic jack is used to raise the front of a van. The diameters of master and slave cylinders are 10 mm and 60 mm respectively. An effort of 250 N is required. Calculate the load on the jack.

15 (Figure 15.18) The body of a tipping truck is raised by two hydraulic rams as shown. The mass of the body is 1·25 t. Calculate:
   a) The force in each ram.
   b) The pressure in the fluid.

16 Air is compressed in a cylinder from atmospheric pressure to 0·4 MN/m², at constant temperature. If the final volume of air is 1·5 m³, calculate the volume of air required at atmospheric pressure.

17 The volume of an oxygen cylinder is 0·035 m³. At the beginning of a welding operation the pressure gauge reads 8·9 MN/m² and at the end of the operation it reads 7·4 MN/m². Determine the volume of oxygen used at atmospheric pressure.

18 A volume of 2·6 m³ of a gas at a gauge pressure of 0·2 MN/m² is compressed at constant temperature. What will be the gauge reading when the volume is 1·3 m³?

19 A gas cylinder of volume 0·25 m³ is at a gauge pressure of 5·9 MN/m². If 10 m³ are drawn off at atmospheric pressure, calculate the final pressure in the cylinder.

20 A tank of volume 4 m³ contains a gas at a pressure of 0·2 MN/m² (absolute). A further volume of 6 m³ of the gas at a pressure of 0·4 MN/m² (absolute) is pumped into the tank. Calculate the final pressure in the tank.

Appendices

1 Coefficient of Friction

Material	μ
metal on metal (dry)	0·15–0·6
metal on metal (lubricated)	0·02–0·1
metal on wood	0·2–0·6
metal on leather	0·3–0·6
metal on concrete	0·3–0·7
rubber on concrete	0·6–0·9
ball bearings	0·001–0·002

2 Mass Densities

Material	$\rho - \text{kg/m}^3$ all values $\times 10^3$
fresh water	1
salt water	1·02
mercury	13·6
carbon tetrachloride	1·6
steel (mild)	7·7
cast iron	7·25
concrete	2·2

3 Some Properties of Metals (average values)

Material	E GN/m^2	U.T.S. MN/m^2	U.C.S. MN/m^2	U.S.S. MN/m^2
mild steel	202	450	450	370
cast iron	110	150	750	140
brass	100	125	125	140
copper	95	140	140	185
aluminium	70	95	95	–

Physical quantity	Symbol	Unit	Abbreviation
length	l	metre	m
mass	m	kilogramme	kg
time	t	second	s
plane angle	α, β, θ	degree	°
angle of friction	ϕ	degree	°
area	A, a	square metre	m^2
volume	V	cubic metre	m^3
linear displacement	s	metre	m
velocity	v, u	metre per second	m/s
acceleration	a	metre per second squared	m/s^2
frequency of rotation	n	revolutions per second	rev/s
density	ρ	kilogramme per cubic metre	kg/m^3
force	F	newton	N
moment	M	newton metre	Nm
pressure	P	newton per metre squared	N/m^2
work	W	joule	J
energy	KE PE	joule	J
power	P	watt	W
stress—direct	σ	newton per square metre	N/m^2
stress—shear	τ	newton per square metre	N/m^2
torque	T	newton metre	Nm
modulus of elasticity	E	newton per square metre	N/m^2
coefficient of friction	μ	ratio	
efficiency	η	ratio	
strain—direct	ε	ratio	
specific gravity	s	ratio	

CONSTANTS   Standard gravitational acceleration $g = 9 \cdot 81 \text{ m/s}^2$

Atmospheric pressure $= 0 \cdot 1 \text{ MN/m}^2$

$\pi = 3 \cdot 14$

5 The Greek Alphabet

A	α	Alpha	I	i	Iota	P	ρ	Rho
B	β	Beta	K	κ	Kappa	Σ	σ	Sigma
Γ	γ	Gamma	Λ	λ	Lambda	T	τ	Tau
Δ	δ	Delta	M	μ	Mu	Y	υ	Upsilon
E	ε	Epsilon	N	ν	Nu	Φ	ϕ	Phi
Z	ζ	Zeta	Ξ	ξ	Xi	X	χ	Chi
H	η	Eta	O	o	Omicron	Ψ	ψ	Psi
Θ	θ	Theta	Π	π	Pi	Ω	ω	Omega

Answers

Chapters 2 & 3: Forces

1 a) 165 N   b) 165 N ⦛82°
2 17·1 N, 13·3 N
3 48·2°, 35·4°
4 346 N, 200 N
5 a) 150 N   b) 20·5°
6 5·66 kN
7 a) 196 N   b) 340 N ⦛60°
8 a) 75·2 kN   b) 101·2 kN
9 a)(i) 312 N   (ii) 624 N
     (iii) 540 N
   b)(i) 488 N   (ii) 384 N
     (iii) 540 N
10 a) 201 N   b) 214 N
11 a) R_A431 N ⦛30°
       R_B528 N 45°⦛
    b) 249 N; R_R533 N
12 a) 52 N   b) 56·4 N ⦛37°
13 a) A 718 N; B 508 N
    b) 732 mm from L.H.E.
14 a) 83·9 N   b) 83·9 N 20°⦛
    c) 230 N ⦛70°

15 a) 3 m   b) 816 N 74°⦛
    c) 226 N →
16 a) 14·6 kg   b) 155 N 67·3°⦛
17 a) 5·12 kN   b) 3·56 kN 44°⦛
18 a) 5·86 N   b) 4·76 N 65°⦛
19 a) 2·16 kg   b) 29 N 60°⦛
20 a) 1 000 N   b) 2 766 N
    c) 1 148 N ⦛75°
21 a) 219 kg   b) 2 127 N
22 a) 6·32 N
    b) α = 110°; θ = 40°
23 A, 453 N; B, 453 N;
    C, 785 N; D, 800 N;
    E, 790 N; F, 422 N
24 a) 7·02 kN; 1·82 kN
    b) 7·02 kN ⦛
       1·82 kN 30° ⦛60°
25 5·33 kN ⦛11°
26 1 233 N ⦛90°

Chapter 4: Moments

1 a) (i) 625 N   (ii) 875 N ↑   2 520 mm
   b) (i) 2·4 m   (ii) 260 N   3 27 kg

4 a) 350 mm from L.H.E.
   b) 137 N
5 R_A 103 N; R_B 63·8 N
6 R_A 2·4 kN; R_B 5·6 kN
7 R_A 2·46 kN; R_B 2·31 kN
8 a) 1·525 tonne
   b) 3·95 kN ↑
9 226 N
10 a) 13 Nm C.M.;
      19·62 Nm A.C.M.
   b) 15·3 N
   c) 63·6 N ⦨69°
11 25 N
12 375 N
13 40 mm
14 a) (i) 70 N
      (ii) 80·6 N ⦨29·7°
   b) (i) 600 mm
      (ii) 283 N ⦠45°
15 71·4 N; 140 N ↑; 28·6 ↓
16 102 kg; 740 N ↓
17 210 N
18 a) F_1 60 N; F_2 100 N
   b) 111 N   c) 49 N ↑
   d) 10·1 mm

19 a) 7 mm   b) 35 N
   c) 17·5 N
20 a) 800 N   b) 1 200 N ↑
21 312 N
22 24 Nm
23 280 N
24 600 N
25 a) 160 Nm   b) 2 kN
26 a) 5·32 kN   b) 1·82 kN
   c) 600 Nm
27 a) 11·55 kN
   b) 2·3 MN/m²
28 a) A, 11·05 kN;
      B, 13·5 kN
   b) 19·62 kN
29 A, 16·35 kN; B, 18 kN
30 a) 21·4 kN
   b) A, 20·05 kN;
      B, 20·95 kN;
      C, 37·5 kN
31 a) 4 kN   b) 6 kN
   c) C, 8·5 kN;
      D, 17·5 kN
32 a) R_A15·1 kN; R_B9·52 kN
   b) 15·05 kN
   c) 0·5 m to left of YY.

Chapter 5: Centre of Gravity

1 a) $\overline{X}$ 30 mm, $\overline{Y}$ 40 mm
   b) $\overline{Y}$ 100 mm
   c) $\overline{X}$ 0·66 m, $\overline{Y}$ 0·69 m
   d) $\overline{Y}$ 61·4 mm
   e) $\overline{X}$ 0·244 m, $\overline{Y}$ 0·384 m
   f) $\overline{X}$ 108 mm, $\overline{Y}$ 109 mm
   g) $\overline{Y}$ 78·9 mm
   h) $\overline{X}$ 95·6 mm

   i) $\overline{X}$ 50·4 mm, $\overline{Y}$ 25·4 m
   j) $\overline{X}$ 37·2 mm, $\overline{Y}$ 30 mm
   k) $\overline{Y}$ 4·67 cm
2 $\overline{X}$ 0·22 m
3 $\overline{X}$ 6·54 cm
4 D 105 mm
5 $\overline{X}$ 0·92 m
6 $\overline{X}$ 1·67 m, T 26·2 kN

7 $\overline{X}$ 41·6 mm, $\overline{Y}$ 8·75 mm
8 $\overline{X}$ 81·6 mm, $\overline{Y}$ 14·4 mm
9 $\overline{X}$ 43 mm, $\overline{Y}$ 25 mm
10 $\overline{X}$ 52·5 mm, $\overline{Y}$ 12·14 mm
    Tension 2·3 N
11 $\overline{X}$ 1·26 m, $\overline{Z}$ 0·16 m,
    mass $0·824 \times 10^3$ kg

12 $\overline{X}$ 10·1 cm
13 $\overline{X}$ 2·34 m, $\overline{Z}$ 1·27 m,
    $\overline{Y}$ 1·13 m
14 1·32 t
15 a) θ 45°, b) θ 35°
16 0·234 m

Chapter 6: Framed Structures

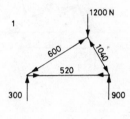

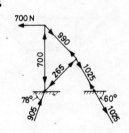

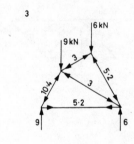

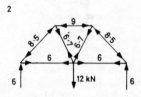

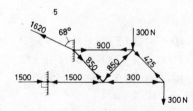

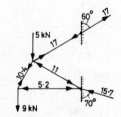

7

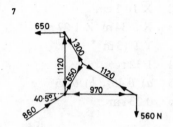

8

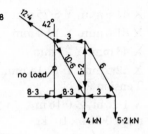

9

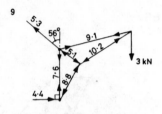

10

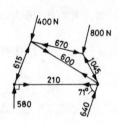

11

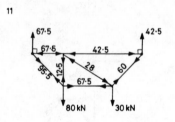

12

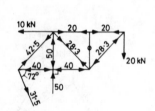

Chapter 7: Friction

1	0·466	9	385 kg
2	40·5 N	10	a) 1·05 kN b) 179 kg
3	a) 0·71 b) 1·81 kN	11	a) 2·57 kN b) 4·16 kN
4	141 N	12	a) 26·5° b) 4·95 kN
5	0·15		c) 1·7 kN
6	1 050 N	13	a) 0·085 b) 4·8°
7	75 kg	14	a) 7·22 kN b) 37·4 kN
8	11·5°	15	a) 166 N b) 1·66 Nm

16   *a)* 40·5 N   *b)* 3·65 Nm
    c) 325 W

17   *a)* 27·6 Nm   *b)* 22·1 N

18   0·3

19   98·5 mm

20   *a)* 437 N   *b)* 19·25 Nm

21   1·67 kN

22   736 N

23   *a)* 85 N   *b)* 59·5 N
    c) 39·2 Nm

24   0·8 kN

25   93 kN

26   440 N

27   1·2 kN/t

28   49

29   *a)* 132 kN   *b)* 99·7 kN

30   *a)* 8·57 kN   *b)* 4·29 kN
    c) 120

31   2·4 kN

Chapter 8: Work Power & Energy

1   5 kJ

2   51·5 kJ

3   150 J

4   33·6 kJ

5   212·5 MJ

6   *a)* 1·25 kJ   *b)* 250 W

7   *a)* 750 N   *b)* 187·5 W

8   4·9 kW

9   2·81 kJ

10   *a)* 53·1 kJ   *b)* 1·8 kW

11   60 kJ

12   *a)* 1 350 J   *b)* 337·5 W

13   1·13 kJ

14   *a)* 4·8 Nm   *b)* 724 W

15   1·95 Nm

16   496 W, 14·9 kJ

17   *a)* 2 m   *b)* 7·5 s

18   2·51 kW

19   *a)* 40 N   *b)* 20 N
    c) 200 mJ   *d)* 200 mJ

20   *a)* 309 J   *b)* 18·7 mm

21   2·45 kW

22   *a)* 40 t   *b)* 2 400 m³
    c) 8·8 MW

23   *a)* 7·84 N, 3·92 N
    b) 0·147 Nm   *c)* 18·5 J
    d) 1·23 W

24   *a)* 5·62 Nm   *b)* 3 kJ
    c) 85 rev

25   1 000 N, 81 J

26   *a)* 220 N   *b)* 3·68 J

27   *a)* 10·3 kJ   *b)* 9·775 kJ
    c) 2·09 kW

28   *a)* 90 N   *b)* 11·25 Nm
    c) 990 W

29   720 kJ

30   *a)* 472 N   *b)* 7·38 MJ

31   *a)* 49 MJ, 64·8 MJ
    b) 817 kW   *c)* 91 km/h

32   707 W

33   230 N, 92 N

34   2·24 kW

35   23·2 rev/s, 15·5 rev/s,
    14·7 rev/s

36   95 N, 42·5 N

Chapter 9: Work Diagrams

1   a) 675 mJ   b) 300 mJ
2   169 J
3   1·95 MJ
4   12 kJ
5   1 510 kJ, 308 kJ
6   86·28 kJ
7   325 J
8   517·5 kJ, 573·8 kJ
9   7·5 MJ
10 2·25 kJ
11 5 mJ

Chapter 10: Machines

1   5
2   MA 5, η 0·83
3   a) 19·2   b) 22   c) 625 J
    d) 715 J   e) 0·875
4   a) 288 J   b) 324 J
    c) 36 J   d) 0·889
5   a) 12·5   b) 43·77 m
    c) 1·313 kJ
6   a) 4   b) 3·34   c) 83·4
    d) 900 J
7   a) 2·13   b) 179 J
    c) 208   d) 0·86
8   a) 1·31 kN   b) 7·12
    c) 735 N
9   a) 441 J   b) 375 J
    c) 0·85   d) 0·102
10 a) 5   b) 0·728
    c) 1·515 kJ
11 a) 5·25
    b) 3 in each block
12 a) 2   b) 278 N
    c) 1·37 kN
13 a) 18   b) 158·5 N
    c) 487 N
14 a) 0·3 m   b) 0·352

15 a) 300   b) 6 kN
16 a) 377   b) 23·8 kN
    c) 16·8 kN   d) 210 Nm
17 a) 15   b) 22·5 mm
    c) 3·6 J   d) 3·15 J
    e) 87·5%
18 a) 1 000   b) 200   c) 0·2
19 a) 57·2   b) 17·8
    c) 0·312
20 P = 0·14 W + 15, 92 N
21 a) P = 0·15 W + 10
    b) 44 N   c) 0·73
22 a) P = 0·072 W + 24
    b) 11   c) 0·22   d) 0·278
23 a) P = 0·28 W + 12
    b) 0·72   c) 0·678
24 a) 10   b) 84·7   c) 14 J
    d) 25·5 N
25 75 teeth, 120 mm
26 0·55, 0·667, 0·834, 1·2,
    1·5, 1·8
27 45 rev/min
28 a) 16·5   b) 1·15 kN
29 a) 8·33   b) 9·6 kJ
    c) 10·65 kJ   d) 1·05 kJ

30 *a*) 18·2 *b*) 190 kg
31 *a*) 1·08 kN *b*) 1·95 kN
    c) 11 *d*) 151
    e) overhaul

32 *a*) 5·6 *b*) 0·795 rev/s
    c) 4·46 rev/s *d*) 250 W
    e) 294 W

Chapter 11: Engines

1 0.75×10^6 N/m^2
2 360 J
3 2·39 kJ
4 3·92 kW
5 12·1 kW
6 1·08 MN/m^2
7 41·8 kW
8 2·34 kW
9 *a*) 0·9 MN/m^2 *b*) 17 kW
10 17·6 kW
11 8 kW
12 352 W
13 82 N
14 0·842
15 *a*) 10·5 kW *b*) 7·8 kW
    c) 74·2%
16 *a*) 25·4 kW *b*) 21 kW
    c) 0·825
17 *a*) 5·5 kW
    b) 6·47 kW, 0·97 kW

Chapter 12: Equations of Motion

1 42 km
2 3 m/s
3 7·5 m/s^2
4 *a*) 187·5 m *b*) 1·67 m/s^2
5 *a*) 2.5×10^3 m/s^2
    b) 20 m/s
6 *a*) 6 m/s^2 *b*) 33·3 s
7 *a*) 75 m *b*) 6 m/s^2
8 *a*) -1.7×10^3 m/s^2
    b) 21·25 mm
9 3·03 s
10 70 s
11 87 m/s, 7 m/s^2
12 *a*) 8 m/s *b*) 40 m
    c) 55 s
13 *a*) 44 m *b*) 14·7
    c) 29·4 m/s
14 21·25 s, 42·5 m/s

Chapter 13: Strength of Materials (1)

1 *a*) (a) 78·63 MN/m^2
       (tension)
    (b) 12·5 MN/m^2
    (compression)

b) 19·55 mm
c) 20 mm × 12 mm

2   144 kN

3   80 kg

4   a) 84·94 MN/m²
    b) 20·74 MN/m²

5   521 kN/m²

6   121 kN/m²

7   19%

8   0·2 × 10⁻³

9   3·75 m

10 a) 50 MN/m²
    b) 0·45 × 10⁻³
    c) 111 GN/m²

11 5·31 mm

12 a) 0·346 mm
    b) 0·239 mm
    c) 0·237 mm

13 a) 20 mm   b) 3·43 m

14 a) 95·5 MN/m²;
       118 MN/m²
    b) 0·261 mm

15 a) 170·2 kN
    b) 1·253 mm

16 a) 40 mm diameter
    b) 0·66 mm
    c) 254·4 MN/m²

17 a) 200 GN/m²
    b) i) 72 kN ii) 0·9 mm

18 200 GN/m²

19 a) 440 MN/m²
    b) 200 MN/m²
    c) 17·3% d) 54·6%

20 a) 112 kN (proportional
       limit); 115 kN (yield
       point) b) 207 GN/m²
    c) 462 MN/m²

Chapter 14: Strength of Materials (2)

1   100 MN/m²

2   140 MN/m²

3   a) 56 kN b) 63·4 kN
    c) 52·4 kN

4   5

5   a) 31·9 MN/m²
    b) 65 MN/m²

6   20·6 mm

7   9 mm

8   4

9   63·7 MN/m²

10 a) 31·4 kN b) 6
    c) 35·7 MN/m²

11 a) 31·8 MN/m²
    b) 44·2 MN/m²

12 a) 40 kN
    b) 78·6 MN/m²
    c) 62·5 MN/m²

13 24·5 mm

14 a) 31·4 kN b) 30 kN
    c) 100 MN/m²;
       150 MN/m²

15 a) 44·8 kN
    b) 112 MN/m²
    c) 25·6 kN

16 a) 8·53 mm b) 5
    c) 30·9%

17 a) 84·9 MN/m²
    b) 73·55 MN/m²
    c) 100 MN/m²
    d) 88·9 MN/m²

e) 4·1 (shear); 4·55
    (direct) f) 5·12

18 73·7 MN/m^2

19 4

20 a) 4 kN
    b) 71·7 MN/m^2
    c) Yes (S.W.S. limit
        72 MN/m^2)
    d) 7·14 mm

Chapter 15: Fluids

1   0·102 MN/m^2, 10·4 m

2   a) 196 × 10^3 N/m^2
    b) 296 × 10^3 N/m^2

3   434 N

4   a) 65 mm b) 114 mm

5   18·55 N

6   0·038 m

7   1 050 m^3

8   0·667 m

9   103·3 kN/m^2

10 3·44 m

11 588·5 N/m^2

12 14·75 × 10^3 N/m^2

13 a) 1·27 × 10^6 N/m^2
    b) 144 N

14 9 kN

15 4·42 kN, 2·25 MN/m^2

16 6 m^3

17 0·53 m^3

18 0·5 MN/m^2

19 2 MN/m^2

20 0·8 MN/m^2

Index